Beginning
iOS 14 & Swift 5
App Development

Greg Lim

Table of Contents

Preface

About this book

In this book, we take you on a fun, hands-on and pragmatic journey to learning iOS 14 application development using Swift 5. You'll start building your first iOS app within minutes. Every section is written in a bite-sized manner and straight to the point as I don't want to waste your time (and most certainly mine) on the content you don't need. In the end, you will have the skills to create an app and submit it to the app store.

In the course of this book, we will cover:
- Chapter 1: Introduction
- Chapter 2: Body Mass Index Calculator
- Chapter 3: Quotes App Using TableView
- Chapter 4: To Do List App Using TableView
- Chapter 5: Persistent Data Using Core Data
- Chapter 6: To Do List with Images & Swipe Delete
- Chapter 7: Connecting to an API: Cryptocurrency Price Tracker
- Chapter 8: Machine Learning with Core ML
- Chapter 9: Augmented Reality with ARKit
- Chapter 10: Publishing Our App on AppStore
- Chapter 11: SwiftUI
- Chapter 12: Widgets
- Chapter 13: App Clips
- Chapter 14: Dark Mode
- Chapter 15: Porting your iOS App to the Mac with Mac Catalyst
- Chapter 16: In-App Purchases

The goal of this book is to teach you iOS development in a manageable way without overwhelming you. We focus only on the essentials and cover the material in a hands-on practice manner for you to code along.

Requirements

No previous knowledge of iOS development required, but you should have basic programming knowledge.

Getting Book Updates

To receive updated versions of the book, subscribe to our mailing list by sending a mail to support@i-ducate.com. I try to update my books to use the latest version of software, libraries and will update the codes/content in this book. So do subscribe to my list to receive updated copies!

Contact and Code Examples

Contact me at support@i-ducate.com to obtain the source codes used in this book. Comments or questions concerning this book can also be directed to the same.

Chapter 1: Introduction

Welcome to Beginning iOS 14 and Swift 5 App Development! I'm Greg and I'm so excited that you decided to come along for this. With this book, you will go from absolute beginner to having your app submitted to the App Store and along the way, equip yourself with valuable iOS app development skills.

Working Through This Book

This book is purposely broken down into sixteen chapters where the development process of each chapter will center it on different essential iOS topics. The book takes a practical hands-on approach to learning through practice. You learn best when you code along with the examples in the book. Along the way, if you encounter any problems, do drop me a mail at support@i-ducate.com where I will try to answer your query.

Get a Mac

Before we proceed on, you will need to have a Mac running on at least macOS version of 10.15.4 (Catalina) to run Xcode 12.

If you do not yet have a Mac, the cheapest option is to get a Mac Mini and if you have a higher budget, get a higher model or iMac with more processing power. You might have heard of the option to run Mac on Windows machines for iOS development, but I do not recommend it. Unexpected problems will arise in development and publishing to the App store that can be avoided by just using a Mac. If you are serious about developing iOS apps and publishing them on the App Store, getting a Mac is a worthwhile investment.

Downloading Xcode

Next, there is an essential piece of software you need to have on your computer before we can move forward. It's called Xcode and is an integrated development environment (IDE) provided by Apple to write Swift code and make iOS apps. It includes the code editor, graphical user interface editor, debugging tools, an iPhone/iPad simulator (to test our apps without real devices) and much more. Let's go ahead to get it downloaded before proceeding.

Download the latest version of Xcode 12 (at time of writing) from the Mac App Store (fig. 1.1).

Xcode
Developer Tools
Apple

3.0 ★★★☆☆   No. 1
191 Ratings   Developer Tools

What's New

• This update fixes an issue that could cause Xcode to crash when viewing documentation

Xcode 12 includes Swift 5.3 and SDKs for iOS 14, iPadOS 14, tvOS 14, watchOS 7, and macOS Catalina

Figure 1.1

You will need an Apple ID to login and download apps from the Mac App store. If you do not already have one, go ahead and sign up for an account (https://appleid.apple.com/account). You will also need an Apple ID to be able to deploy your app to a real iPhone/iPad device for testing.

The installation of Xcode might require you to update your version of MacOS. At this book's time of writing, the MacOS required is Catalina version 10.15.4.

Installing Xcode

Just like any other Mac App, Mac App store will take care of the downloading and installation of Xcode for you. Do note that installation of Xcode 12 requires 20-30 GB of space available for the installation to proceed and installation does take quite some time. Once the installation is complete, you should see the Xcode icon on your computer.

Swift and Xcode

I'm going to be introducing you to two terms that you're going to encounter throughout this book. One of those is Swift and the other one is Xcode. Swift is the programming language we use to make iPhone apps. Swift came out in 2014. Previous to that, the programming language used to make iPhone apps was Objective C. But Objective C was complicated. Many developers new to the space of iOS development found that it was hard to read and write. Swift then was introduced. Swift is specifically designed with beginners in mind and even experienced programmers think of Swift as a really clean and beautiful language.

Xcode is the program that allows us to make iPhone apps. We're going to type Swift into Xcode and also use Xcode for designing the visual side of our app like where do we want a button, what color do we want it to be, where do we want to place our table view, etc.

So throughout this book, these are the two skills that we will be improving upon step by step.

This course was written for a beginner in iOS development. So if you have some iOS development experience you're going to feel pretty familiar with what's going on. It will also be best if you have some basic programming experience. But if you do not have it, it's alright as well as I will try my best to explain certain programming concepts.

Xcode Walkthrough

Now in this section, I want you to become acquainted with Xcode. Go ahead and open Xcode.

At the time of writing, this book uses Xcode 12. But make sure you're using the latest official version of Xcode from the Mac App Store.

In the 'Welcome to Xcode' screen (fig. 1.2), you can choose to either get started with a playground which is a great way to explore the Swift language. The next option is creating a new Xcode project where you create an app for iPhone, iPad, Mac, Apple Watch or Apple TV.

Figure 1.2

You also have a third option to clone an existing project but we will not be covering this option in this book.

For now, let's go ahead and create a new project. When you do so, it's going to bring up a page (fig. 1.3) that asks what kind of project do you want to make, whether iOS, watchOS, tvOS, macOS or Multiplatform.

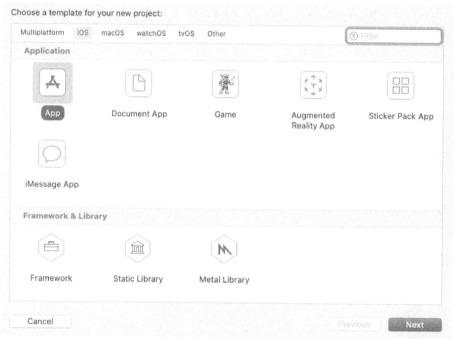

Figure 1.3

iOS includes apps for the iPhone, iPad, iPod Touch. *watchOS* is for Apple watch apps. *tvOS* is for Apple TV apps. *macOS* is for Mac apps on the desktop and *Multiplatform* is if you want to make an app that works across multiple platforms. For us, we will be focusing on *iOS* apps.

For an iOS app, there are lots of different templates that you can start with. The templates help you get started with some boilerplate code. For us, we want the *App* option. This is essentially the blank starting point for almost every app that we're going to make. So let's go ahead and double click on that.

You will then have to input the below fields for your project (fig. 1.4):

Product Name: (as this is our first project, we will name it *HelloWorld*)

Team:

Organization Identifier: (normally the reverse of your website e.g. com.iducate.calculator. If you do not have a website, com.firstname.lastname will do fine)

Interface: select *Storyboard*

Life Cycle: select UIKit App Delegate

Language: select *Swift*

For *Use Core Data, Include Tests*: leave all the boxes unchecked

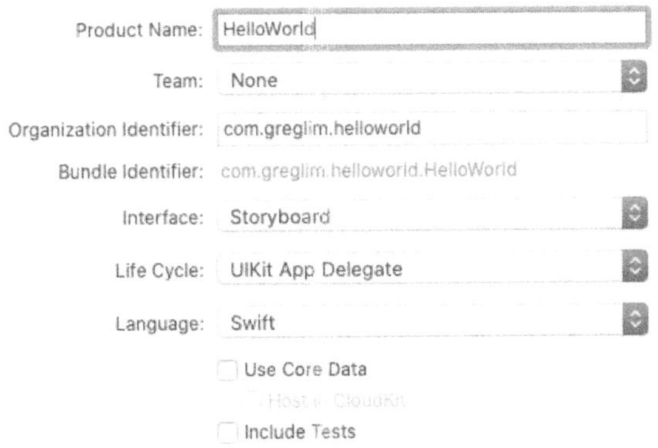

Product Name:	HelloWorld
Team:	None
Organization Identifier:	com.greglim.helloworld
Bundle Identifier:	com.greglim.helloworld.HelloWorld
Interface:	Storyboard
Life Cycle:	UIKit App Delegate
Language:	Swift
	☐ Use Core Data
	☐ Host in CloudKit
	☐ Include Tests

Figure 1.4

*Note: In the User Interface field is an option to select SwiftUI or Storyboard. SwiftUI is a new way of implementing user interfaces introduced in Xcode 11 and iOS 13. It will be the best way to make iOS apps eventually but there are certain things that it can't handle yet. If you try to make a professional app with SwiftUI, you find yourself going back to storyboard elements and it can get messy. So we will select *Storyboard* which had been the established way of designing interfaces. We will introduce SwiftUI in chapter eleven as well as use it to make Widgets in chapter twelve.

Go ahead and fill in the fields. You can change the field values later in your project, so don't worry if you have inputted a wrong value.

When you have the fields filled up, hit the *Next* button. It's going to ask you where you want to save this new project. I'm going to put ours under 'Documents'.

There will also be a checkbox to 'Create Git repository on my Mac'. This will make a GitHub repository for your app which helps you save different versions of your app and if you want to collaborate with people. Git is outside the scope of this book but just go ahead and leave this checked.

You can see that a new folder has been added to our desktop called *HelloWorld*. On the left side of Xcode, you can see the folder-file structure of the project (fig. 1.5).

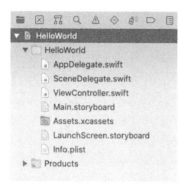

Figure 1.5

We have our *HelloWorld* folder at the root and another *HelloWorld* folder with more files inside of that. Xcode will take the name of your project and put that in a folder which consists of another folder with that same name. It will also have a project file *HelloWorld.xcodeproj* which is the project file to open our project. If we close out Xcode and wanted to open up your project to work on it again, double click on *HelloWorld.xcodeproj* and it will open up Xcode with all your project folders and files.

These files make up our app. A major one here is *Main.storyboard*. This is where all the visual part of our code happens. For example, whenever we want to add a button or label to our app, it's all going to happen inside of *Main.storyboard*.

There are some swift files here like *ViewController.swift* and *AppDelegate.swift* where we're going to be typing out some code to make things happen. We will cover more on them later in this book.

And then we have folders like *Assets.xcassets* to store resources that our app uses. For example, images, sounds, fonts, and videos. The apps in this book will have *Assets.xcassets* mainly storing images.

Info.plist (Information Property List) contains metadata settings for your app e.g. app name, version, and other fields that we have entered in our initial project setup form.

Lastly, if you click back on the root project file, you can see the General settings for your app (fig. 1.6).

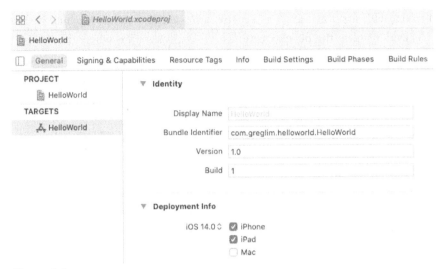

Figure 1.6

This is where previously you had filled in the fields at the beginning of the project. You can come back here and change it.

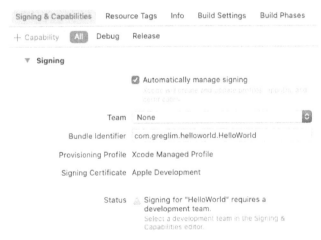

Figure 1.7

Under the 'Signing & Capabilities' tab (fig. 1.7) is also where we had to specify the 'Team' field. Again, you can come back here and change it. For example, we have not specified *Team*. Here, you can go back and change when you want to submit your app to the App Store.

The two main files we will work with for now are *ViewController.swift* and *Main.storyboard* which Xcode created for us by default. As evident from the file extensions, *ViewController.swift* is where the programming code of our app resides and *Main.storyboard* is the visual part of our app. We will later show how to connect the visual part of our app to the Swift code file.

To begin adding visuals, click on *Main.storyboard* where you will be brought to the View Controller scene (fig. 1.7b). You can think of a View Controller as a representation of a screen in your app.

Figure 1.7b

Now click on [+] on the top right-hand corner and the list of visual objects will appear (fig. 1.8).

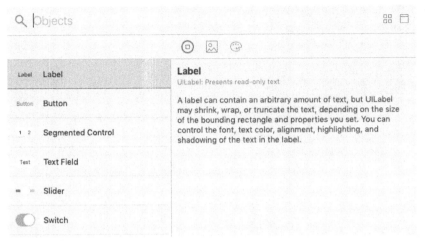

Figure 1.8

You can then search for the kind of visual that you want by typing it into the search field (fig. 1.9).

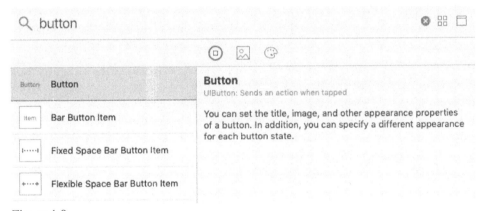

Figure 1.9

For our app, search for a label drag and it onto the View Controller in the storyboard (fig. 1.10, 1.11). Labels are used to represent texts.

Figure 1.10

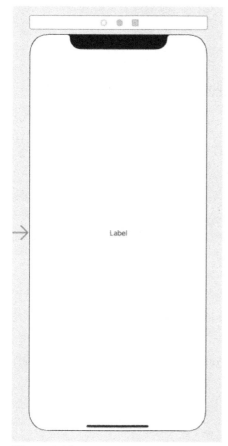

Figure 1.11

You can test to see how your app will look different on different devices, e.g. iPhone 11, iPhone 11 Pro

Max, iPad by selecting 'Device' at the bottom of the storyboard (fig. 1.12)

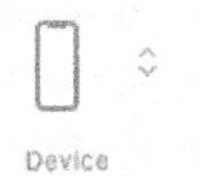

Device

Figure 1.12

Let's change back to the iPhone 11 device.

You will realize that for different devices, the app looks different. That is if we try to center our label on an iPhone 11 and when we change the display to an iPhone 8 plus or an iPhone 4s, the centering goes off. To fix this, we will need to add *constraints*.

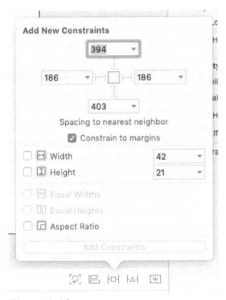

Figure 1.13

To add new constraints, select the label and click on the ⊡ icon (fig. 1.13). We then specify 2 constraints of 10 from the left and 10 from the right after which select 'Add 2 Constraints' (fig. 1.14).

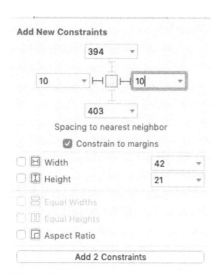

Figure 1.14

What this means is that we constraint the label to begin with a margin of 10 from the left and end with a margin of 10 from the right.

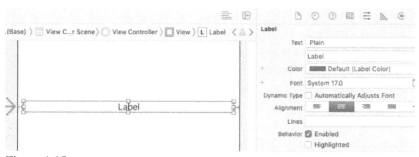

Figure 1.15

To center the text, select the label and in 'Attributes Inspector', select 'Center' for alignment (fig. 1.15). Thanks to constraints, our label and text will now be centered regardless of which device we select now.

Try it yourself

As a mini challenge, in the label, add a constraint of margin 400 from the top yourself.

To change the text of the label, we just have to double click on the label and specify the text we want. But suppose we want to programmatically change the text value of our label from code, how do we do so? For that, we will need to use Outlets.

Outlets

Outlets are what connect the visual part (storyboard) to the code part (Swift files) of our app. We use an outlet in our Swift file to get access to a label in the storyboard. To begin creating an outlet for our app, it is useful to switch to the 'Split screen' mode. To do, select the View Controller in the storyboard and click on the ☰ 'Adjust Editor Options' icon (fig. 1.16a).

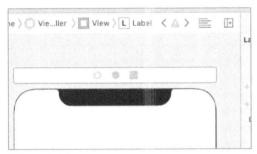

Figure 1.16a

Select 'Assistant' (fig. 1.16b).

Figure 1.16b

At this point, there should be a split screen showing your storyboard on the top and the code file for 'ViewController.swift' on the bottom (fig. 1.16c). (*Note: depending in your monitor size, it might show storyboard on the left and code file on the right)

'ViewController.swift' shows up because it is the code file that is connected to the storyboard on the left.

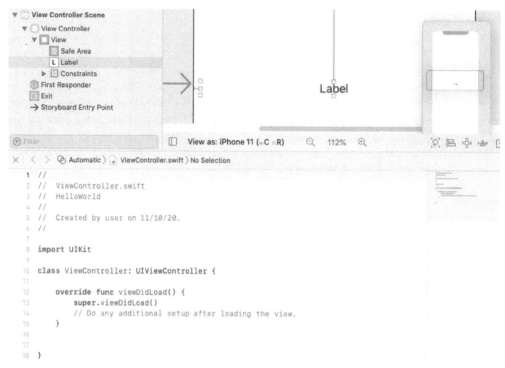

Figure 1.16c

Once in split screen, select the label, hold down 'Control' key and click drag to the *ViewController.swift* file just above the *viewDidLoad()* function. You will be prompted 'Insert Outlet or Outlet collection' (fig. 1.17).

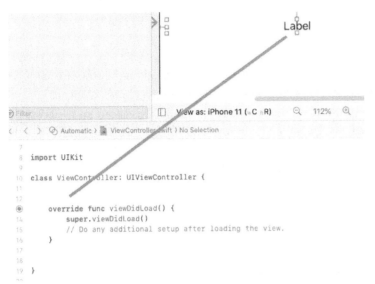

Figure 1.17

When you let go of the mouse, you will be prompted to enter the 'Name' of the label outlet(fig. 1.18).

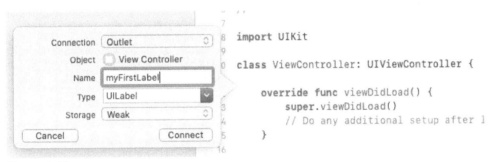

Figure 1.18

Name it 'myFirstLabel', click 'Connect' and the below code in bold will be generated for you.

```
import UIKit

class ViewController: UIViewController {

    @IBOutlet weak var myFirstLabel: UILabel!

    override func viewDidLoad() {
        super.viewDidLoad()
```

```
        }
}
```

Xcode has created for us a variable called *myFirstLabel* that refers to the label on the storyboard.

With this, we have established access to the label from *ViewController.swift*. This allows us to access the label and control attributes of the label. For example to edit text, add the line below in bold:

```
class ViewController: UIViewController {

    @IBOutlet weak var myFirstLabel: UILabel!

    override func viewDidLoad() {
        super.viewDidLoad()
        myFirstLabel.text = "Hello World!"
    }
}
```

We will now run our app to see the changes.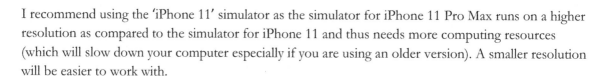

I recommend using the 'iPhone 11' simulator as the simulator for iPhone 11 Pro Max runs on a higher resolution as compared to the simulator for iPhone 11 and thus needs more computing resources (which will slow down your computer especially if you are using an older version). A smaller resolution will be easier to work with.

Now to begin running our app, click the ▶ icon from the panel on the top left of Xcode.

The simulator should start running and soon display your app (fig. 1.19).

Figure 1.19

This flow of adding a visual control in the storyboard and programmatically controlling it inside of a Swift file via an outlet is an important concept that will be used very often. So make sure that you understand it well!

Buttons

Other than labels, buttons are another visual control that you see in almost all apps. Users click on buttons which in turn lead to actions calling specific code.

In the same project, go back to the storyboard and drag a button into the View Controller. Like the label, set the left and right margin constraints to be the same as the labels (fig. 1.20).

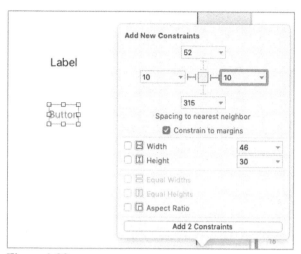

Figure 1.20

Although our left and right margin constraints have been added, the vertical distance between our label and button has yet to be fixed. This might cause different vertical distances on different devices. To fix a constraint for the vertical spacing, select the button, hold down the 'Control' key, click drag onto the label and select 'Vertical Spacing' (fig. 1.21).

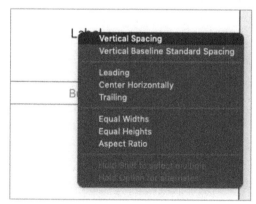

Figure 1.21

'Vertical Spacing' means that there should always have the same amount of vertical space between these two visual controls across different devices.

Button Actions

To create an action for our button, similar to how we created an outlet for the label, select the button, hold the 'Control' key down, and click drag over to the Swift code just after the *viewDidLoad()* method (fig. 1.22).

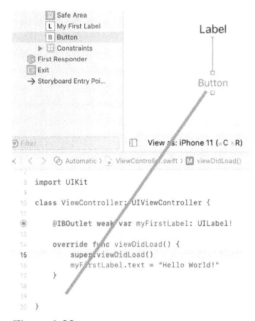

Figure 1.22

A popup with *connection* 'Action' will appear. (fig. 1.23) Name the action 'buttonTapped' and click *connect*.

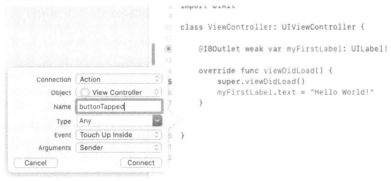

Figure 1.23

You will then have the action function *buttonTapped* created for you as shown below in bold:

```
import UIKit

class ViewController: UIViewController {

    @IBOutlet weak var myFirstLabel: UILabel!

    override func viewDidLoad() {
        super.viewDidLoad()
        myFirstLabel.text = "Hello World!"
    }
    @IBAction func buttonTapped(_ sender: Any) {
    }
}
```

To test if our action indeed gets called when the button is tapped, we can add a *print* statement in the action:

```
@IBAction func buttonTapped(_ sender: Any) {
    print("button tapped")
}
```

Now when you run the app in the simulator and you click on the button, you should have "button tapped" printed in the console log.

Now that we know how to create outlets to refer to visual controls from code and buttons to call actions, let's modify our app to create a counter app. A counter app is useful especially when you organize events and want to count the number of attendees. You might have seen a physical one before.

Try it Yourself!

Now let's create a counter app! In our counter app, there should be a label which shows how many times the button has been clicked. And each time the button is clicked, the count in the label should be incremented by one. We will then have another button which resets the count to zero. You will also have a variable to store how many times the button has been clicked. Before I walk through you the process of creating the app, why not try creating the app on your own? And only if you run into problems, that you could read through my walkthrough. As renowned educator Howard Hendricks says, "Maximum learning involves maximum involvement." So go ahead and try creating the counter app on your own!

Have you tried it yet? If you have and succeeded, congratulations! For those who have run into problems, let's now walk through the solution together.

Solution

First, we will need a variable to store how many times the button has been clicked. Declare a variable *count* and initialize it to 0 as shown below:

```
class ViewController: UIViewController {

    var count = 0

    @IBOutlet weak var myFirstLabel: UILabel!
        ...
```

Second, we increment *count* by one each time *buttonTapped* is called as shown below. We also convert *count* to a String using *String(count)* and assign it to *myFirstLabel*'s *text* attribute. The *text* attribute of visual controls is what gets displayed in the user interface.

```
@IBAction func buttonTapped(_ sender: Any) {
    count = count + 1
    myFirstLabel.text = String(count)
}
```

Our app currently shows "Hello" when it first starts. We have to instead show the count when it starts as shown below:

```
override func viewDidLoad() {
    super.viewDidLoad()
    //myFirstLabel.text = "Hello World!"
    myFirstLabel.text = String(count)
}
```

You can also change the text of the button to "Count". The app should look something like (fig. 1.24):

Figure 1.24

And each time you click on the 'Count' button, the counter increments by one!

The final code should look like:

```
import UIKit

class ViewController: UIViewController {

    var count = 0

    @IBOutlet weak var myFirstLabel: UILabel!
```

```
override func viewDidLoad() {
    super.viewDidLoad()
    myFirstLabel.text = String(count)
}
@IBAction func buttonTapped(_ sender: Any) {
    count = count + 1
    myFirstLabel.text = String(count)
}
}
```

Try it yourself:

Now, can you implement the 'Reset' button functionality which resets count to zero when pressed? Remember to create an action for the reset button and in it set count to 0. Contact support@i-ducate.com if you face any issues.

Summary

We began the journey on iOS14 and Swift App Development to go from absolute beginner to having our app submitted to the App Store. We went through the steps of downloading and installing Xcode, having a walkthrough of the Xcode interface, using Xcode to create an App. We learned about adding constraints to visual elements to ensure that the app looks the same for different devices of different sizes.

Chapter 2: Body Mass Index Calculator

In this chapter, we will be making a Body Mass Index (BMI) Calculator app. If you are unfamiliar with BMI, it helps to see if we are at a healthy weight. To work out your BMI:
- divide your weight in kilograms (kg) by your height in meters (m)
- then divide the answer by your height again to get your BMI

For example: If you weigh 70kg and you're 1.75m tall, divide 70 by 1.75 – the answer is 40 then divide 40 by 1.75 – Your BMI is 22.9.

We then have the following classifications:
Underweight: Your BMI is less than 18.5.
Healthy weight: Your BMI is 18.5 to 24.9.
Overweight: Your BMI is 25 to 29.9.
Obese: Your BMI is 30 or higher.

We will be creating a new project for our BMI Calculator. In particular, we will learn about Text fields to get input from users. So close the current one, and once again go through the steps needed to create a new project: 'Create a New Xcode Project', select 'App', name it 'BMI Calculator', uncheck all the boxes as we won't be using them yet.

In *Main.storyboard*, drag two textfields from the object library for our weight and height inputs with the weight textbox on top of the height textbox. Do you remember how to center the two textfields and also how to create vertical spacing between them across multiple devices?

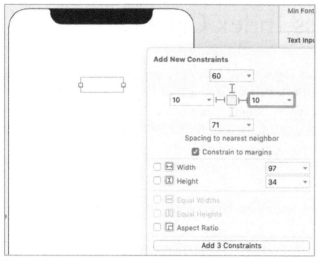

Figure 1

In case you forgot or need a refresher, for each text field, select it and in 'Add New Constraints', add the left and right margin constraints of 10 (fig. 1). Do this both for the weight and height textfield. Additionally for the weight textfield, add a top margin constraint of 60.

Hold 'Control' and click drag from one text field to the other to fix the vertical spacing (fig. 2).

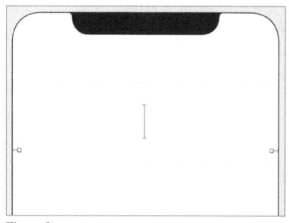

Figure 2

We then add placeholder texts for each of our textfields to prompt the user on what to fill in. For each textfield, go to 'Attributes Inspector' and in 'PlaceHolder', type 'Enter Weight' for the weight textfield (fig. 3) and 'Enter Height' for the height textfield.

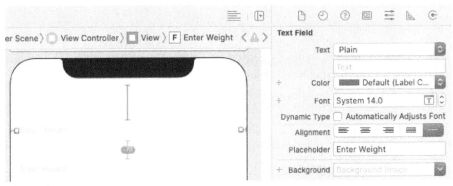

Figure 3

We should also be using the Number Pad keyboard type instead of the generic keyboard for the user to enter height and weight since it's in numbers. To do so, select each textfield, in 'Attributes Inspector', 'Keyboard Type', select 'Number Pad' (fig 4).

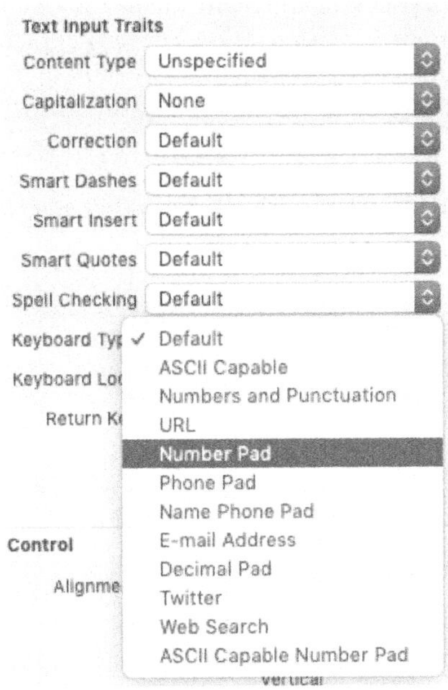

Figure 4

You can see that there are many other kinds of keyboard types available for different kinds of expected inputs. Go and experiment with different various keyboards on your own.

TextField Outlets

Remember how we used outlets to programmatically control labels from our Swift code? We need to do the same and create outlets for our textfields to retrieve the input from it. So proceed to create outlets for the two textfields. Try doing it on your own!

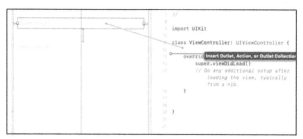

Figure 5

Remember to select textfield, hold the 'Control' key down and drag click to the Swift file just above *viewDidLoad*(fig. 5) to create the Outlet. Name your outlets *weightTextField* and *heightTextField* as shown:

```
class ViewController: UIViewController {

    @IBOutlet weak var weightTextField: UITextField!
    @IBOutlet weak var heightTextField: UITextField!

    override func viewDidLoad() {
        super.viewDidLoad()
    }
}
```

Next, drag a button into the View Controller, center it by creating left/right margin constraints for it and create vertical spacing for it below the height textfield (fig. 6).

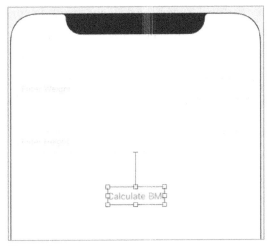

Figure 6

Create an action for the button by holding 'Control' and click drag to the Swift file just below *viewDidLoad()*. Name it *buttonTapped* and fill it with the following code:

```
@IBAction func buttonTapped(_ sender: Any) {
    let weight = Double(weightTextField.text!)!
    let height = Double(heightTextField.text!)!

    let bmi = weight/(height * height)
    print(bmi)
}
```

In the above code, we assign *weight* and *height* with the user inputted value from the textfield. We convert their inputted value from String to Double using *Double(...)* because we will be executing mathematical operations with it. You might be asking, what is the use of the exclamation mark "!" at the end of the line? We will be explaining this in a section later entitled, "Optionals".

Now when you run your app in the simulator, fill in your weight in kilograms and height in meters (e.g. weight: 80, height: 1.7) and when you click 'Calculate BMI', you have your BMI index printed in the console! How did you fare? My BMI isn't too good. I had better exercise more!

Now we don't want to just print the calculated BMI to the console where a user can't see it. So we will be displaying the BMI in a label. Drag a label below the 'Calculate BMI' button.

Once again, create left/right margin constraints of 10 for the label, center the text, and create a vertical spacing from the button on top. Create an outlet 'bmiLabel' for it. And in *buttonTapped*, add the following:

```
@IBAction func buttonTapped(_ sender: Any) {
    let weight = Double(weightTextField.text!)!
    let height = Double(heightTextField.text!)!

    let bmi = weight/(height * height)

    bmiLabel.text = String(bmi)
}
```

Now the calculated BMI should appear in the label (fig. 7).

Figure 7

We should however only display the result to 1 decimal place. To format our result to 1 d.p., we do the following:

```
bmiLabel.text = String(format: "%.1f", bmi)
```

and our BMI should now display a nicely formatted to 1 d.p. result.

BMI Classification

Next, we want to append to the BMI classification: "BMI: 29.4, Overweight"

We have to implement the below logic in our Swift code:
Underweight: Your BMI is less than 18.5.
Healthy weight: Your BMI is 18.5 to 24.9.
Overweight: Your BMI is 25 to 29.9.
Obese: Your BMI is 30 or higher.

To do so, add the following code:

```swift
@IBAction func buttonTapped(_ sender: Any) {
    let weight = Double(weightTextField.text!)!
    let height = Double(heightTextField.text!)!

    let bmi = weight/(height * height)

    var classification:String

    if bmi < 18.5{
        classification = "Underweight"
    }
    else if bmi < 24.9{
        classification = "Healthy weight"
    }
    else if bmi < 29.9{
        classification = "Overweight"
    }
    else{
        classification = "Obese"
    }

    let formattedBMI = String(format: "%.1f", bmi)
    bmiLabel.text = "BMI: \(formattedBMI), \(classification)"
}
```

Code Explanation

```
var classification:String
```

We first declare a variable *classification* to store the classification result e.g. Underweight.

```
if bmi < 18.5{
    classification = "Underweight"
}
else if bmi < 24.9{
    classification = "Healthy weight"
}
else if bmi < 29.9{
    classification = "Overweight"
}
else{
    classification = "Obese"
}
```

We next then have a series of *if-else* statements to determine the classification as per the BMI classification logic.

```
let formattedBMI = String(format: "%.1f", bmi)
bmiLabel.text = "BMI: \(formattedBMI), \(classification)"
```

We assign the formatted BMI to 1 d.p. first to a variable *formattedBMI*. Next, we use Swift string interpolation "\(string)" to append text before and after the calculated result.

When you run your app now, the BMI classification should be nicely appended to the calculated BMI (fig. 8).

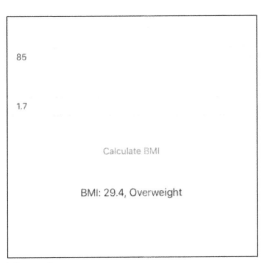

Figure 8

var vs *let*

Now is a good time to revisit our code and look into when we should use *var* and when to use *let*. For example in our code, we declare *weight*, *height* and *bmi* using *let*.

```
let weight = Double(weightTextField.text!)!
let height = Double(heightTextField.text!)!
let bmi = weight/(height * height)

var classification:String

if bmi < 18.5{
    classification = "Underweight"
}
    ...
```

But we have declared *classification* using *var*.

The *let* keyword defines a constant, that is, the value won't be changed afterward. In our case, *weight*, *height* after being inputted by the user and *bmi* after being calculated won't be changed, thus we use *let*. We can of course use *var* for these variables and the code will still work. But Xcode will prompt you a warning to use *let* instead (fig. 9).

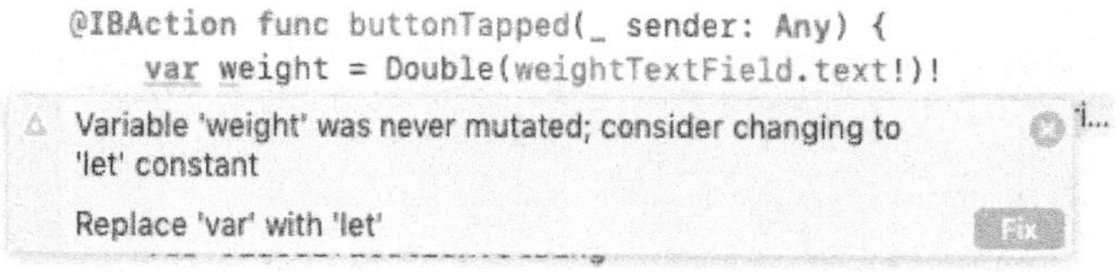

```
@IBAction func buttonTapped(_ sender: Any) {
    var weight = Double(weightTextField.text!)!
```

△ Variable 'weight' was never mutated; consider changing to   ⊗ ¹...
    'let' constant

    Replace 'var' with 'let'   Fix

Figure 9

It is good practice to use *let* for variables which value never changes. *var* however defines an ordinary variable whose value changes during runtime. For example, *classification* changes during runtime.

Optionals

Earlier, we had the below code with an exclamation mark at the end:

```
let weight = Double(weightTextField.text!)!
let height = Double(heightTextField.text!)!
```

So why do we need it?

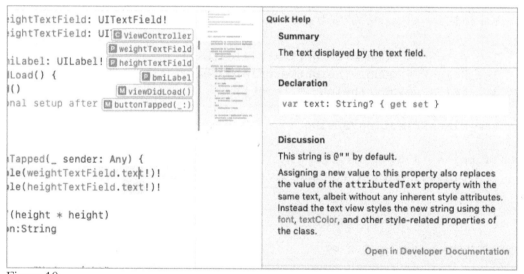

Figure 10

If you place your cursor at *weightTextField.text* and look under the 'Quick Help', you will see that *.text* returns "String?" which is a String optional (fig. 10). Any variable type with a '?' as suffix will make an existing type an optional. This is to specify that either this variable has a value, or it is nil. For example, *String?* would either have a string value or nil. *Int?* would either have an integer value or nil.

You might ask, why should *weightTextField.text* return an optional? Shouldn't it just return a string? Now when a user does not enter any value into a textfield, what should *weightTextField.text* return? Therefore, *weightTextField.text* returns an optional, that is, it returns the string value entered by the user and returns nil when user does not enter anything in the textfield.

To unwrap an optional, that is, we say get the value directly, we add "!" as a suffix to a variable. But as you might have guessed, it is quite dangerous to do so especially when there's the possibility that our variable indeed has a nil value. In fact, if you run the BMI calculator app and leave the field blank and calculate BMI, the app will crash (fig. 11).

```
@IBAction func buttonTapped(_ sender: Any) {
    let weight = Double(weightTextField.text!)!
    let height  =  Thread 1: Fatal error: Unexpectedly found nil while unwrapping an
                   Optional value
    let bmi = weignt/(neignt * neignt)
```

Figure 11

To avoid our app crashing when a textfield is left blank, we should be doing:

```
var weight: Double = 0

if weightTextField.text != nil{
    if Double(weightTextField.text!) != nil{
        weight = Double(weightTextField.text!)!
    }
}
```

That is, we have to first check each optional that if it's not nil, then proceed to unwrap it. If it's nil, assign weight as 0.

A better way to unwrap the optional would be to use '*if let*'. '*if let*' first checks if an optional variable contains an actual value and bind the non-optional form to a temporary variable. This is the safe way to "unwrap" an optional or in other words, access the value contained in the optional.

```
var weight: Double = 0;

if let weightText = weightTextField.text {
    if let weightDouble = Double(weightText) {
        weight = weightDouble
    }
}
```

In the above code, we are saying, only move forward if *weightTextField.text* is not nil and in such a case, assign the value string to temporary variable *weightText* (*weightText* does not exist outside the scope). And in the inner *if*, only move forward when *Double(weightText)* is not nil (i.e. it's a valid double numerical value) and assign the value double to temporary variable *weightDouble*. Finally, assign *weightDouble* to *weight*.

Summary

We went through fundamental iOS development building blocks in the form of outlets, actions and using them to control visual elements such as buttons and textfields. Applying that knowledge, we went to create our first app, a counter app and also a Body Mass Index calculator. We also introduce optionals in Swift. In the next chapter, we will explore how to use tableviews in our app.

Chapter 3: Quotes App Using TableView

In this chapter, we will be building a quotes app where you have a list of quotes in a tableview (fig. 1). When you tap on a quote, you get to a quote details page which shows the whole quote plus the author of the quote.

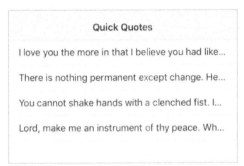

Figure 1

The concept we will be learning here is the tableview which is an important UI element that most apps have. Tableviews are commonly used in apps to list out information. An example is Whatsapp where we can scroll through different chatgroups (fig. 2).

Figure 2

Other examples include 'Settings' and 'Contacts' (fig. 3).

Carrier
11:10 AM
Groups
Contacts
+
Search
A
John Appleseed
B
Kate Bell
H
Anna Haro
Daniel Higgins Jr.
T
David Taylor
Z
Hank M. Zakroff
A B C D E F G H I J K L M N O P Q R S T U V W X Y Z #

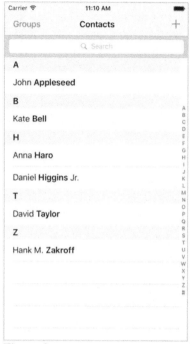

Figure 3

First, begin a new 'App' project in Xcode, call it *QuickQuotes*. Under 'Language' choose 'Swift', under 'User Interface' choose 'Storyboard' and leave the below checkboxes unchecked (fig. 4).

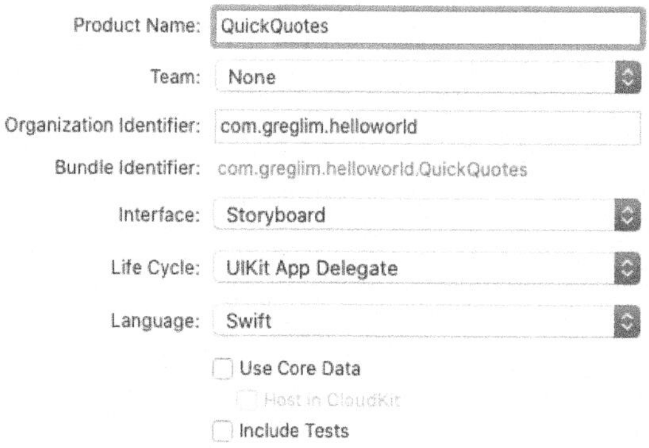

Product Name: QuickQuotes
Team: None
Organization Identifier: com.greglim.helloworld
Bundle Identifier: com.greglim.helloworld.QuickQuotes
Interface: Storyboard
Life Cycle: UIKit App Delegate
Language: Swift
Use Core Data
Host in CloudKit
Include Tests

Figure 4

We will visually add a tableview in the storyboard. As you already have seen, our project when created comes with a default blank View Controller. As mentioned, a view controller represents a screen. When we move from one screen to another, we are switching between view controllers.

We will delete the default view controller to start afresh. Select the view controller by selecting it under the 'View Controller Scene' dropdown (fig. 5) and press 'Delete'.

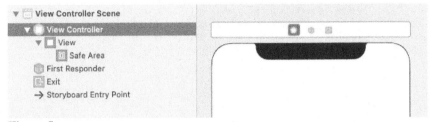

Figure 5

Remember that there is also a linked *ViewController.swift* file? We will have to delete that as well. Select the file, choose 'Delete' and 'Move it to Trash' (fig. 6).

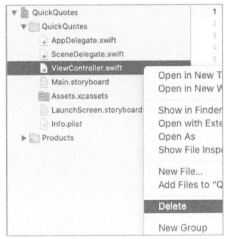

Figure 6

Back in the storyboard, click the ＋ icon and search for 'Table View Controller' (fig. 7) and drag it into the storyboard (fig. 8).

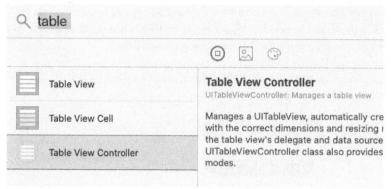

Figure 7

Figure 8

A table view controller is simply a view controller with a table view in it. Now we want this table view controller to be the initial view controller when the app first starts. To do, select the table view controller by clicking on its top panel, go to its 'Attributes Inspector' and check the 'Is Initial View Controller' checkbox (fig. 9).

Figure 9

An arrow will now appear to the left of the table view controller. The arrow tells us where the app starts with if we have multiple view controllers. If we run the app now, we can see the table view controller displaying rows of little lines which you can scroll up and down (fig. 10).

Figure 10

Just as we previously have a *ViewController.swift* to control our View Controller in the storyboard, we will also need to add a new Swift file to control our Table View Controller.

To add a new Swift file, go to 'File > New > File...' (fig. 11).

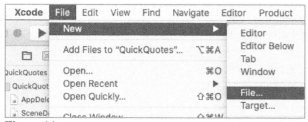

Figure 11

Select 'Cocoa Touch Class' and click 'Next' (fig. 12).

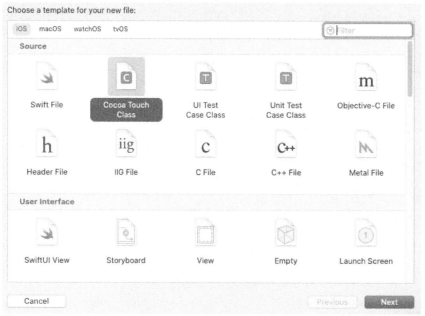

Figure 12

The reason we chose a Cocoa Touch Class and not a Swift file for the table view is that Cocoa Touch Class imports UIKit which provides the classes, properties and methods associated with typical UI view controllers.

Under 'Subclass of', select 'UITableViewController' and name the class 'QuoteTableViewController'. Also ensure that 'Also create XIB file' is unchecked, since we have already created it visually in the storyboard (fig. 13).

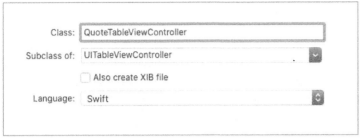

Figure 13

Click 'Next', 'Create' and the file will be added to our project (fig. 14).

Figure 14

Next, we have to connect our tableview controller to the newly added Swift file. Select the table view controller by clicking on the yellow circle icon, go to identity inspector ⊞ , and in *Class*, type in 'QuoteTableViewController' (fig. 15).

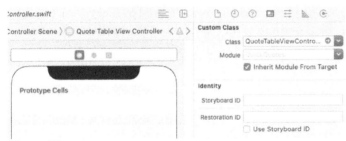

Figure 15

With this, we have connected our table view controller in the storyboard to our *QuoteTableViewController.swift* file. This is a very important step!

Populating our Table View

We will now go through how to begin populating our table view. If you look at *QuoteTableViewController.swift*, you might be overwhelmed with a lot of code. But you will notice that majority of the code are commented out and only *viewDidLoad()*, *numberOfSections* and *tableView* are uncommented out. If we look at the comments for the code, they actually give a good explanation for the purpose of each function.

QuoteTableViewController.swift being a subclass of *UITableViewController*, means that we are provided with a class that has a default implementation of *UITableViewDelegate* and *UITableViewDataSource* protocols.

For example, the most important functions for UITableViewDataSource are:
```
numberOfSections(in:)
tableView(_:numberOfRowsInSection:)
tableView(_:cellForRowAt:)
```

That is, these methods will be called to obtain data used to display in the table view, i.e. number of sections, number of rows in each section and what to present for each cell.

The relevant functions for *UITableViewDelegate* are:
```
tableView(_:didSelectRowAt:)
```
That is, we manage what will happen when a row is selected by implementing the *didSelectRowAt* method.

In *numberOfSections*, we are supposed to return the number of sections that our table view controller will have.

```
override func numberOfSections(in tableView: UITableView) -> Int {
    //#warning Incomplete implementation, return the number of sections
    return 1
}
```

In our case, we do not yet have different sections, so we just return 1 for now. But say for example in our contact lists, contacts might be categorized according to alphabets, and thus we would have 26 sections each representing an alphabet.

```
override func tableView(_ tableView: UITableView, numberOfRowsInSection
    section: Int) -> Int {
    // #warning Incomplete implementation, return the number of rows
    return 0
}
```

tableView(numberOfRowsInSection) as the name implies specifies how many rows are there in a section. For example, say we want 10 rows, we return 10:

```
override func tableView(_ tableView: UITableView, numberOfRowsInSection
  section: Int) -> Int {
    return 10
}
```

Next, you will see *tableView(cellForRowAt)* commented. *cellForRowAt* is called for each cell row. A cell represents a single row and in *cellForRowAt*, we define how a cell is displayed. Uncomment the code for *cellForRowAt* and add in the following code:

```
override func tableView(_ tableView: UITableView, cellForRowAt indexPath:
  IndexPath) -> UITableViewCell {
    let cell = UITableViewCell()
    cell.textLabel?.text = "My Quote"
    return cell
}
```

You will notice in the method header -> *UITableViewCell* that the method returns a *UITableViewCell* which represents a single row. Thus, we define a variable of type *UITableViewCell* and then set its *textLabel*'s text property to "MyQuote". Lastly we return the cell. When we run our app, we get 'My Quote' populated in each cell (fig. 16):

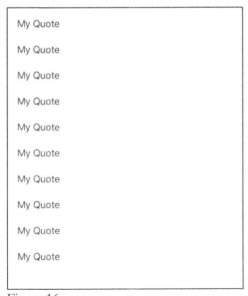

Figure 16

This is the basic idea to get a tableview populated. We specify the number of rows in our table view in *numberOfRowsInSection* and the details of each row with *cellForRowAt*. For now, we are displaying the same text for each row. We will later show how to display distinct values for each row by reading from an array.

It is worth mentioning that Apple has made displaying of table views very efficient. For example, you can try returning 10,000 rows in *numberOfRowsInSection* and when you run your app, it still scrolls fine without any lag! Try even 1,000,000!

Populating Rows from an Array

We will now show how to populate each row with values from an array. First, declare an array of strings in QuoteTableViewController:

```
class QuoteTableViewController: UITableViewController {

    var quotes = [
        "I love you the more …",
        "There is nothing permanent …",
        "You cannot shake hands …",
        "Lord, make me an instrument…"
    ]
    ...
```

For space's sake, I have shortened the quotes in this book. But of course you can fill in your own quotes.

Next in *numberOfRowsInSection*, return *quotes.count*:

```
    override func tableView(_ tableView: UITableView, numberOfRowsInSection
section: Int) -> Int {
        return quotes.count
    }
```

You can see that we no longer return a hardcoded value of 10. We should rather be returning the size of the array to displaying the corresponding number of rows. And then in *cellForRowAt*, we assign the textLabel with *quotes[indexPath.row]*:

```
    override func tableView(_ tableView: UITableView, cellForRowAt indexPath:
IndexPath) -> UITableViewCell {
        let cell = UITableViewCell()
        cell.textLabel?.text = quotes[indexPath.row]

        return cell
    }
```

indexPath.row contains which row *cellForRowAt* is currently called for. For example, for the first row, *indexPath.row* will return 0. For the second row, it will return 1. Thus, if you add print(indexPath.row) to cellForRowAt, you can see the console printing:

0

1

2

3

We thus can use the array index to refer to each corresponding element in the array. If you run your app now, it should display the quotes in the array (fig. 17).

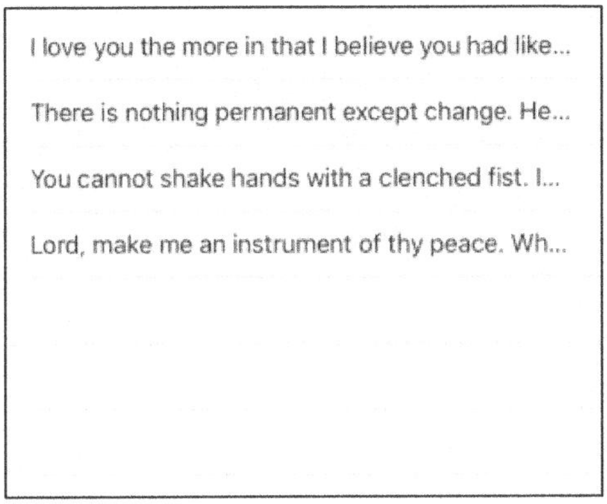

```
I love you the more in that I believe you had like...

There is nothing permanent except change. He...

You cannot shake hands with a clenched fist. I...

Lord, make me an instrument of thy peace. Wh...
```

Figure 17

Take note that if a string is too long for each cell, it will automatically be shortened with "...".

Segues

Because a quote might be too long to be displayed in a cell, we will next implement a quote details screen to show the entire quote. That is, when a user taps on a quote, we will go to a separate quote details view controller. This is where we need segues. A segue defines a transition between two view controllers in an app's storyboard file. A segue has a starting point like a button, table row or gesture which initiates the segue. The end point of the segue is the view controller we want to display. In our case, the segue we are about to implement starts with a table row and ends in the quote details view controller.

First, select the Table view controller and under 'Editor' > 'Embed In', select 'Navigation Controller' (fig. 18).

Figure 18

You will see that the 'arrow' has moved to the navigation controller which in turn points to our table view controller. Embedding in navigation controller is a common practice by many developers because it provides an easy way to show new view controllers (push) and dismiss child view controller with a back button. It also provides a navigation bar which gives us a place for a title and a tab bar for action buttons.

As you can see, a navigation bar has been added to the top of our table view controller (fig. 19).

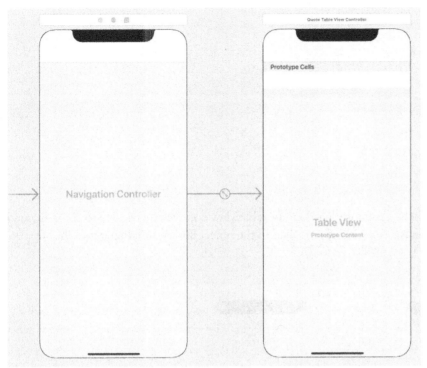

Figure 19

Now select the navigation header bar and under 'Attributes Inspector', fill in 'Title' as 'Quick Quotes' (fig. 20).

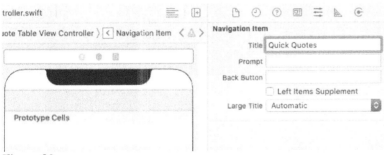

Figure 20

And if you run your app now, you should see the title 'Quick Quotes' displayed on the navigation bar header (fig. 21).

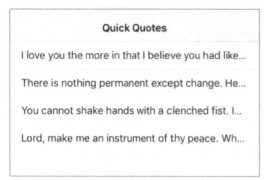

Figure 21

Now we need to add another view controller that our table view controller navigates to show the quotes detail. In the storyboard, search for 'View Controller' in the object library and drag it into the storyboard to the right of the table view controller (fig. 22).

Figure 22

We next need to connect our table view controller to the new view controller using segues. To do so, click on the left circle of the table view controller to select the whole controller, hold down the 'Control' key and click drag on to the new view controller (which should light up in blue, fig. 23).

Fig. 23

When you let go, you will have different segues to choose from (fig. 24).

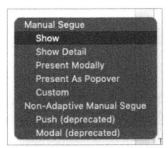

Figure 24

Choose 'Show' and there will be a connection added. For the *Show* segues, the destination view controller is pushed onto the navigation stack, moving the source view controller out of the way (destination slides overtop from right to left).

The top bar in the new view controller will now show 'Quick Quotes' with a back symbol (fig. 25).

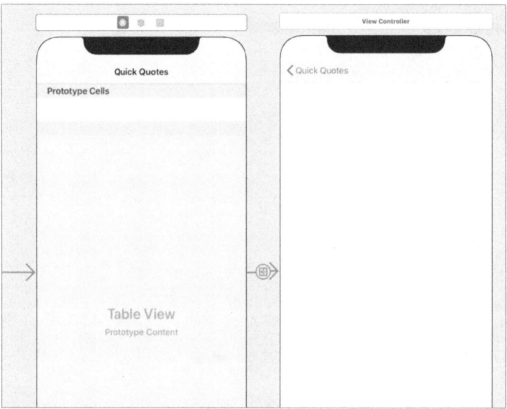

Figure 25

(*The other segues are used more for iPad applications for example, 'Show Detail' is used in a *UISplitViewController* in an iPad app, 'Popover Presentation' - when run on iPad, the destination appears in a small popover, and tapping anywhere outside of this popover will dismiss it.)

Now, we have to give our segue a name to identify it. To do so, select the segue and in 'Attributes Inspector' > 'Identifier', fill in 'moveToQuoteDetail'. The identifier should be a name that describes the segue (fig. 26).

Figure 26

Next, we want code to run when a quote is tapped to show the appropriate quote in the details page. To do, in *QuoteTableViewController.swift*, add the *didSelectRowAt* function. When you start typing it in, Xcode's intelligence filler should automatically provide you with the function which you can choose from (fig. 27).

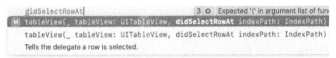

Figure 27

didSelectRowAt will be called when a user taps on one of our table view cells. Now fill in the below codes into *didSelectRowAt*:

```
    override func tableView(_ tableView: UITableView, didSelectRowAt indexPath:
IndexPath) {
        let selectedQuote = quotes[indexPath.row]
        performSegue(withIdentifier: "moveToQuoteDetail", sender: selectedQuote)
    }
```

Code Explanation

```
let selectedQuote = quotes[indexPath.row]
```

First, we have a variable *selectedQuote* which stores the selected cell user has tapped on. Similar to *cellForRowAt*, we access the *quotes* array with *indexPath.row* which tell us which row is being tapped and assign that array element to *selectedQuote*.

```
performSegue(withIdentifier: "moveToQuoteDetail", sender: selectedQuote)
```

Next, we use *performSegue* to move to the next view controller. *performSegue* expects two arguments. The first is the identifier of the segue which we want to perform, in our case 'moveToQuoteDetail'. The second argument is the data we want to pass to the destination which in our case is *selectedQuote*.

Now, we have sent *selectedQuote* over, but how does the new view controller receive that data? To do so, we first need to have a Swift code file for the view controller. This is something we had done before. Can you try adding it on your own? (Don't worry, I will still walk you through it)

Go to 'File' > 'New File', select 'Cocoa Touch Class'. In 'Subclass of', select 'UIViewController' and name the class 'QuoteDetailViewController'.

Connect the class 'QuoteDetailViewController' to the View controller in the storyboard, select the View controller in the storyboard (by clicking on the top left button) and under 'Identity Inspector' > 'Custom class', 'Class', fill in *QuoteDetailViewController* (fig. 28). The View controller will now be linked to *QuoteDetailViewController.swift*.

Figure 28

Prepare

Back in *QuoteTableViewController.swift*, we will write a new function that will handle passing of the quote from the current table view controller to the *QuoteDetailViewController*.

In *QuoteTableViewController.swift*, you will see a commented-out function called *prepare*. Uncomment the function and fill in the following codes:

```
override func prepare(for segue: UIStoryboardSegue, sender: Any?) {
    // Get the new view controller using segue.destination.
    // Pass the selected object to the new view controller.
  if let quoteViewController = segue.destination as? QuoteDetailViewController{
      if let selectedQuote = sender as? String {
         quoteViewController.title = selectedQuote
      }
   }
}
```

Code Explanation

The *prepare* function is called right before we move on to the next view controller. As the default comment says, "In a storyboard-based application, you will often want to do a little preparation before navigation" which is the purpose of *prepare*.

```
if let quoteViewController = segue.destination as? QuoteDetailViewController{
    ...
}
```

We check if the segue's destination is of type QuoteDetailViewController (using *segue.destination as? QuoteDetailViewController*). If so, assign the view controller reference to *quoteViewController*. This thus gives us a handle to our destination view controller.

```
if let quoteViewController = segue.destination as? QuoteDetailViewController{
    if let selectedQuote = sender as? String {
        quoteViewController.title = selectedQuote
    }
}
```

Next, we check if the object to be sent type is of type string using "as? String" and let *selectedQuote* hold the data. We then assign *selectedQuote* to the destination's *title* attribute.

Now when we run our app and click on a cell, we will be brought to the new view controller with its title displaying the quote we clicked! (fig. 29)

Figure 29

Showing the Quotes

Showing the quote in the header of the details page is obviously not ideal. We get only a truncated quote. We will need a big label to show the entire quote. In the next few steps, we will drag a label into our details view controller, connect our label to code using an outlet so that we can assign the quote text to it.

First, let's add the label to our details view controller. In storyboard, drag a label to the details view controller. We want our label to fill up the entire screen. To do so, select our label and in 'Add New Constraints', specify 0 for all the margins (left, right, top, bottom) and click 'Add 4 Constraints' (fig. 30).

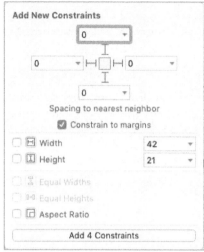

Figure 30

Your label should now fill the entire screen (fig. 31).

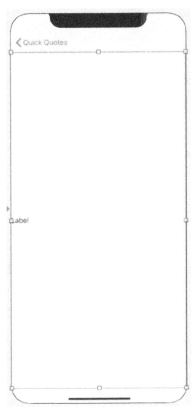

Figure 31

Connect the label to *QuoteDetailViewController.swift* to create an outlet for it and name it 'quoteLabel'. You should be able to do this on your own now.

But how do we assign our quote text to the label? Previously, we assigned the quote text to the *title* property of QuoteDetailViewController. Now, we will declare a variable to hold the text as shown below:

```
import UIKit

class QuoteDetailViewController: UIViewController {

    var quote = ""

    @IBOutlet weak var quoteLabel: UILabel!

    override func viewDidLoad() {
        super.viewDidLoad()
```

63

```
        quoteLabel.text = quote
    }
}
```

Notice also that we have added **quoteLabel.text = quote** in *viewDidLoad()* to have the label show the quote when the screen loads. Currently *var quote* is still an empty string. To have it contain the quote text, make the following change in *QuoteTableViewController.swift*:

```
override func prepare(for segue: UIStoryboardSegue, sender: Any?) {
    if let quoteViewController = segue.destination as? QuoteDetailViewController{
        if let selectedQuote = sender as? String {
            quoteViewController.quote = selectedQuote
        }
    }
}
```

When you run your app now and click on a quote, it should display something like (fig. 32):

Figure 32

Our quote is still truncated as it displays in one single line. We should change this to display it in as many lines as required. To do so, select the label and under 'Attributes Inspector', 'Lines', specify '0' (fig. 33). '0' means that we want as many lines as required to display the text.

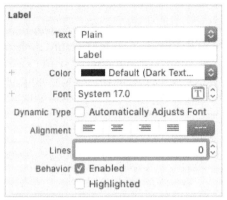

Figure 33

To make the text more presentable, you can also specify 'Alignment' for the label text. In my case, I have specified 'Justified' alignment (fig. 34).

Figure 34

Congratulations! You have your first working tableview app which navigates to a details controller!

Summary

In this chapter, we have built a quotes app which lists quotes in a tableview and upon tapping on a quote, get directed to a quote details page. We learned the concept of the tableview, an important UI element to list out information. We covered how to connect our table view controller in the storyboard to a Swift file, how to populate the table view with values from an array and how to transit from the main page to a details page using segues. In the next chapter, we will reinforce on concepts covered in this chapter as well as learn how to add items to a table view.

Chapter 4: To Do List App Using TableView

In this chapter, we will be building the classic 'To Do List' app (fig. 1). Whether it's a shopping list, errands list, so on, the classic 'To Do List' app always serves as a foundational example for mobile app development. In the process of building the app, we will also revisit and cement the concept of table view controllers and navigation controllers.

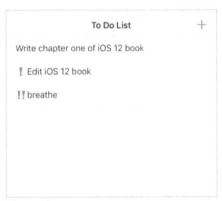

Figure 1

So go ahead and create a new 'App' project in Xcode. This time however, check the 'Core Data' checkbox and name the project 'TodoList' (fig. 2).

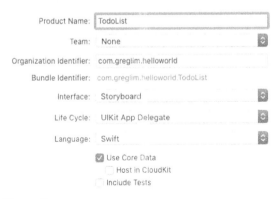

Figure 2

Listing To-Do Items in a TableView

Like the Quotes app, we will be using a table view to list our to do items. The following step should be familiar to you. Like before, delete the default view controller in the storyboard and also its

ViewController.swift file for us to start anew.

Now, drag in a navigation controller which has a default connected table view controller (fig. 3).

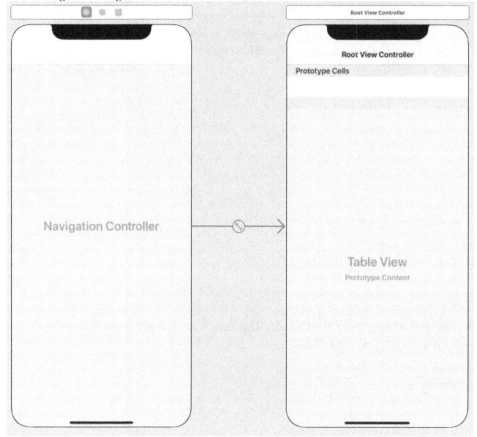

Figure 3

Make sure that in 'Navigation Controller', 'Attributes Inspector', under 'View Controller', 'Is Initial View Controller' checkbox is checked.

Rename the title of table view controller to 'To Do List' (fig. 4).

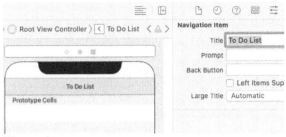

Figure 4

Next, we will need to have a class to represent a 'to-do' item. Add a new class Swift file by going to:

'File' > 'New' > 'File…', and choose 'Swift File' (fig. 5, do not choose Cocoa Touch class!).

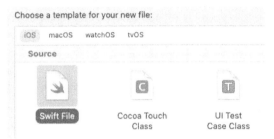

Figure 5

Name it *ToDo*, and in the project folder, you will see the file *ToDo.swift* created. In *ToDo.swift* add in the lines:

```
import Foundation

class ToDo{
    var name = ""
    var priority = 0
}
```

Our *ToDo* class will have a name for the to-do item and a *priority* property of type integer. If the priority value is 2, we will show two !! exclamation marks as prefix to the todo name. If priority value is 1, we show one exclamation mark. Else, we don't add any exclamation as prefix. This is similar to what Apple's Reminder app displays (fig. 6):

Figure 6

Next, we will create the Swift code file for our table view controller. Go to 'File' > 'New' > 'File...', select 'Cocoa Touch class', select 'Subclass of' as *UITableViewController* and name it *ToDoTableViewController.swift*. As covered in the previous chapter, being a subclass of *UITableViewController*, Apple will have implemented many functions vital to a table view controller for us.

At the top of the class, add the below code:

```
var toDos: [ToDo] = []
```

This declares and initializes an empty array called *toDos*. It will be containing *ToDo* type objects.

We previously returned 1 for '*numberOfSections*' since we don't have sections. Alternatively, we can just delete the function. So go ahead and delete it.

Next, in '*numberOfRowsInSection*', add:

```
override func tableView(_ tableView: UITableView, numberOfRowsInSection
```

70

```
    section: Int) -> Int {
        return toDos.count
    }
```

We will then proceed to create some dummy to-do items to fill our array:

```
    override func viewDidLoad() {
        super.viewDidLoad()

        let toDo1 = ToDo()
        toDo1.name = "Write chapter one of iOS 14 book"
        toDo1.priority = 0
        let toDo2 = ToDo()
        toDo2.name = "Edit iOS 14 book"
        toDo2.priority = 1

        toDos = [toDo1,toDo2]    // fills up the array
    }
```

Next uncomment *cellForRowAt* and fill in the below codes:

```
    override func tableView(_ tableView: UITableView, cellForRowAt
indexPath: IndexPath) -> UITableViewCell {
        let cell = UITableViewCell()
        let selectedToDo = toDos[indexPath.row]

        if selectedToDo.priority == 1{
            cell.textLabel?.text = "!" + selectedToDo.name
        }
        else if selectedToDo.priority == 2{
            cell.textLabel?.text = "!!" + selectedToDo.name
        }
        else{
            cell.textLabel?.text = selectedToDo.name
        }

        return cell
    }
```

In the above, we first create a *UITableViewCell* object. We then retrieve the specified to-do item from the array using *indexPath.row*. Based on the to-do item's priority value, we then add as prefix, an exclamation emoji. If priority is 1, add one exclamation. If 2, add two. Else, don't do anything.

Although in the above code, I added '!', we can actually add the emoji into the code!

In Mac OS Mojave and above, to add emoji, press 'Control', 'Command' and 'Space' together to bring up the emoji keyboard. Search for 'exclamation' and you should find the exclamation emoji (fig. 7).

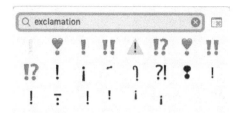

Figure 7

Now, we will connect *ToDoTableViewController.swift* to our storyboard table view controller. Select the table view controller in storyboard, in 'Identity Inspector', under 'Custom Class', in 'Class', fill in 'ToDoTableViewController' (fig. 8).

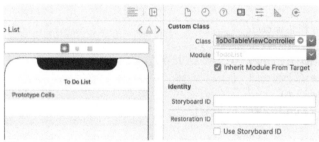

Figure 8

Adding To-Do Items

Next, we will implement adding to-do items. Currently, our to-do items are hard-coded. In the storyboard, drag and drop a 'bar button' on to the top right of the table view controller. Select the newly added bar button and under 'Attributes Inspector', 'System Item', select 'Add' (fig. 9).

Figure 9

You will notice that Xcode gives us a nice '+' icon for our button. There are other options for other functions that we can choose from. For example, 'Edit'. You will however realize that for 'Edit', there are no icons, it displays just the text. So why should we use 'System Item' when we can directly change the text? The good thing about using 'System Item' is that it supports localization. That is, if a user changes the phone language to Chinese for example, 'Edit' will be translated into Chinese.

Now drag a new view controller into the storyboard. This will serve as the Add To-Do Item View Controller.

Next, from *ToDoTableViewController* in the storyboard, hold down 'Control', click drag from the '+' button to the new view controller and in segues, select 'Show'. You will notice a new connection made and a back button has been added in the new view controller (fig. 10).

Figure 10

In the new view controller, we will drag a couple of controls into the view controller so that it will look something like (fig. 11):

Figure 11

So proceed to drag a textfield, a label, a segmented control and a button into the view controller like in figure 11. Make sure that you set margin of 10 from both left/right side and vertical spacings between each of them so that the UI looks the same across multiple devices.

For the textfield, in 'Attributes Inspector', 'Placeholder', fill in 'Describe item'
For the button, change the text to 'Add'.
For the Segmented Control, specify '3' in 'Attributes Inspector', 'Segments' to have three segments (fig. 12).

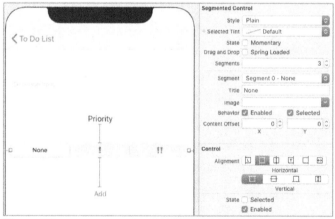

Figure 12

Select each segment from the dropdown, and specify (fig. 13):
- 'None' for segment 0
- '!' for segment 1 (use the emoji keyboard by press 'Ctrl','Cmd', 'Space' together)
- '!!' for segment 2

Figure 13

For segment 0, under 'Behavior', 'Selected', check the 'Selected' checkbox. This selects 'None' by default

Creating Outlets

Now, we need to have the Swift code version of the view controller and in it, create outlets for our textfield, segmented control and action for our button.

By now you should already be familiar with how to create a new view controller swift file and link it to the storyboard. So create a new Cocoa touch class file (subclass of *UIViewController*) and name it *AddToDoViewController.swift*. Once that's created, select the view controller in the storyboard and in 'Identity Inspector', 'Custom Class', 'Class', specify *AddToDoViewController.swift*.

Proceed to create outlets for the textfield and segmented control by holding 'Ctrl' and drag clicking to the code file. I have named them *nameTextField* and *prioritySegment*.

```
@IBOutlet weak var nameTextField: UITextField!
@IBOutlet weak var prioritySegment: UISegmentedControl!
```

Button Action

Also, create the action for the button and name it 'addTapped'. In it, we retrieve the user inputted values to create our *ToDo* object with the following code:

```
@IBAction func addTapped(_ sender: Any) {
    let newToDo = ToDo()
    newToDo.priority = prioritySegment.selectedSegmentIndex
    if let name = nameTextField.text {
        newToDo.name = name
    }
}
```

Code Explanation

```
let newToDo = ToDo()
```

We first create a *ToDo* object called *newToDo*.

```
newToDo.priority = prioritySegment.selectedSegmentIndex
```

We then specify the priority of *newToDo* with the selected segment index of *prioritySegment*. *prioritySegment.selectedSegmentIndex* returns 0 for the first segment, 1 for the second and 2 for the third. That is why we have specified *priority* in *ToDo* as an integer and why previously in the tableview's *cellForRowAt* function that we do:

```
if selectedToDo.priority == 1{
    cell.textLabel?.text = " ! " + selectedToDo.name
}
else if selectedToDo.priority == 2{
    cell.textLabel?.text = "!!" + selectedToDo.name
}
...
```

Next, we retrieve the user inputted values from the *text* property of *nameTextField*.

```
if let name = nameTextField.text {
    newToDo.name = name
}
```

Because *nameTextField.text* returns us a string optional, we use *let name* to contain the string value only when there is a value to proceed on to assign to *newToDo.name*.

Reloading the Table View

We have created our *ToDo* object, but how are we going to pass this newly *ToDo* object to our tableview for it to be displayed? We will need a *reference* to our table view from our *AddToDoViewController*. So how do we create this reference?

We create this reference by going back to the table view controller and implementing it in the *prepare* function. Remember that the *prepare* function is where we do a little preparation before navigation. Because the *prepare* function has a reference to the destination view controller, that is we have a reference from the table view controller to the *add* view controller. We can then access the *add* view controller and create a reference back to the table view controller.

To do so, in *ToDoTableViewController.swift*, uncomment the *prepare* method and fill in the following codes:

```swift
override func prepare(for segue: UIStoryboardSegue, sender: Any?) {
  if let addToDoViewController = segue.destination as? AddToDoViewController{
    addToDoViewController.toDoTableViewController = self
  }
}
```

In the above code, we first check if the segue destination is of *AddToDoViewController* type. If so, we create a variable in *addToDoViewController* called *toDoTableViewController* and assign it with the current table view controller (using *self*). This thus creates a reference from *Add* view controller back to the table controller!

Now of course back in *AddToDoViewController.swift*, we have to add the *toDoTableViewController* reference:

```swift
class AddToDoViewController: UIViewController {
    var toDoTableViewController: ToDoTableViewController?=nil

    @IBOutlet weak var nameTextField: UITextField!
    @IBOutlet weak var prioritySegment: UISegmentedControl!
    override func viewDidLoad() {
        super.viewDidLoad()
    }
      ...
```

(*ToDoTableViewController?=nil* means that *toDoTableViewController* is an optional. That is, it can either have a value or it can be nil)

Next in *addTapped*, fill in the following codes:

```swift
@IBAction func addTapped(_ sender: Any) {
    let newToDo = ToDo()
    newToDo.priority = prioritySegment.selectedSegmentIndex
    if let name = nameTextField.text {
        newToDo.name = name
    }
    toDoTableViewController?.toDos.append(newToDo)
    toDoTableViewController?.tableView.reloadData()
    navigationController?.popViewController(animated: true)
}
```

Code Explanation

Having this reference, in *addTapped*, we now append to *toDos* array in the table view controller with *toDoTableViewController?.toDos.append(newToDo)*.

To update the table view, we call *toDoTableViewController?.tableView.reloadData()*.

After adding the new todo, to go back to the table view, we call *navigationController?.popViewController(animated: true)* to pop the current view controller out of the view stack to reveal the underlying table view controller. This is equivalent to the 'Back' operation. We set *animated:true in popViewController* for a smooth transition between views.

Running your App

Before running your app, make sure that in the Attribute Inspector of Navigation controller, you have checked the 'Is Initial View Controller' checkbox. If you run your app now, you should be able to add to-do items and have your table views showing them (fig. 14).

Figure 14

Passing Data to To-Do Details Page

We will now implement the To Do details page, which is when we select a particular cell in the table view, a view controller that shows the to-do details appear.

In the storyboard, drag in a new (details) view controller. It will be easier to see if you place the new view controller below the table view controller (fig. 15).

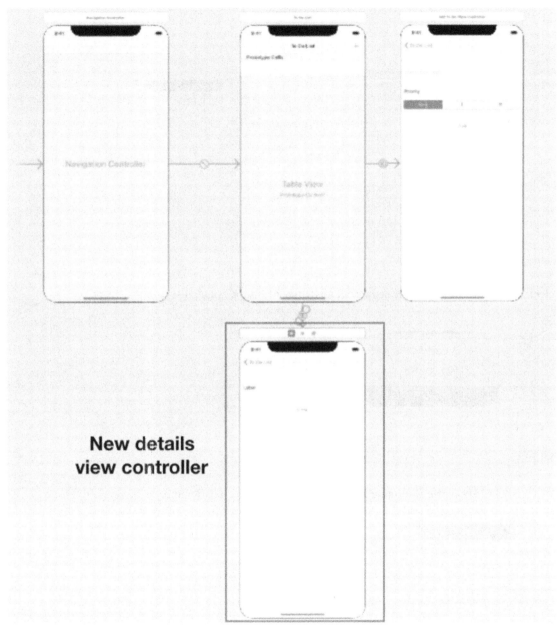

Figure 15

Create a segue from the table view controller to the new details view controller by selecting the cell on the table view, and 'Ctrl' drag click to the details view controller. If you have trouble selecting the table view cell, you can alternatively select 'Table View Cell' from the tree structure and drag click from there (fig. 16):

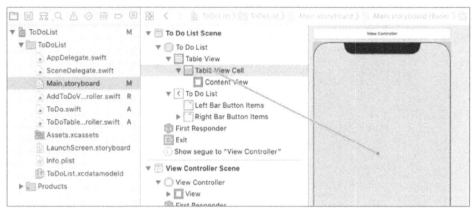

Figure 16

Select 'Show' and next give the segues an identifier of 'moveToDetails' (fig. 17).

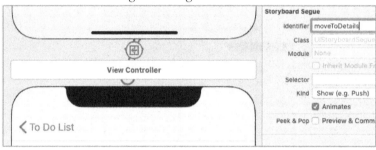

Figure 17

Drag a label to show the to-do text and also a button for the To-Do item to be marked 'Done'. Remember to center and add the necessary margins for your display controls (fig. 18).

Figure 18

Next, create the Swift code file for details view controller but creating a new Cocoa Touch class, subclass of *UIViewController* file and name it ToDo*DetailsViewController.swift*.

Once created, link it to the storyboard by selecting the top left yellow button of the view controller, 'Identity Inspector', 'Custom Class', 'Class' and specifying 'ToDoDetailsViewController' (fig. 19).

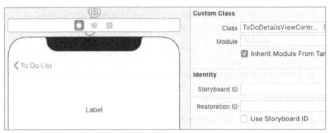

Figure 19

Next, create an outlet for the label and an action for the button as shown below:

```
class ToDoDetailsViewController: UIViewController {

    ...
    @IBOutlet weak var toDoLabel: UILabel!

    ...
    @IBAction func doneTapped(_ sender: Any) {

}
```

Back in *ToDoTableViewController.swift*, implement *didSelectRowAt* function with the below code:

```
override func tableView(_ tableView: UITableView, didSelectRowAt indexPath:
IndexPath) {
      let selectedToDo = toDos[indexPath.row]
      performSegue(withIdentifier: "moveToDetails", sender: selectedToDo)
}
```

Xcode should fill in the function header for you as you begin to type the function name.

Handling Different Segues Destinations

As mentioned previously in our quotes app, *didSelectRowAt* is called when a cell is selected. We refer to the selected cell with *toDos[indexPath.row]* and assign it to *selectedToDo*. We then perform a segue with identifier "*moveToDetails*" with the *sender* object being *selectedToDo*.

Once again, we need to do a little preparation before we perform the navigation in *prepare*. Specifically, we need to pass in the selected to-do over to *ToDoDetailsViewController*. And because *prepare* already has an existing segue preparation to *AddToDoViewController*, we will have another portion of code that performs a different segue preparation based on different view controller destinations. In *ToDoTableViewController.swift*, add in the following code in **bold**:

```
override func prepare(for segue: UIStoryboardSegue, sender: Any?) {
    if let addToDoViewController = segue.destination as? AddToDoViewController{
        addToDoViewController.toDoTableViewController = self
    }

    if let detailsToDoViewController = segue.destination as?
ToDoDetailsViewController{
        if let selectedToDo = sender as? ToDo{
            detailsToDoViewController.toDo = selectedToDo
        }
    }
}
```

As you can see, we have different segue actions based on if the destination is *AddToDoViewController* or *ToDoDetailsViewController*. This is how we can manage different segue actions from a single table (or master) view controller.

For the segue action to *ToDoDetailsViewController*, we assign the *selectedToDo* object and assign it to a variable *toDo* in *ToDoDetailsViewController*. Now we have not added *toDo* in detailsToDoViewController yet, so proceed to add it in:

ToDoDetailsViewController.swift

```
class ToDoDetailsViewController: UIViewController {

    var toDo = ToDo()
        ...
```

With *ToDoDetailsViewController* having the selected to-do object, we can now populate our label with the following codes in *viewDidLoad()*:

```
override func viewDidLoad() {
    super.viewDidLoad()
    if toDo.priority == 1{
        toDoLabel.text = " ! " + toDo.name
    }
    else if toDo.priority == 2{
        toDoLabel.text = "!!" + toDo.name
```

```
        }
        else{
            toDoLabel.text = toDo.name
        }
    }
```

Similar to what we have done in *cellForRowAt* in our table view controller, we append the appropriate number of exclamation emojis based on the priority level.

And when you run your app now and click on a todo item, the todo details screen will show the todo text with the priority.

With this, it should give you some familiarity with how to pass data across different view controllers and segues.

Missing Link

The problem with our app now is that the data is not persistent. If we add new to-do items, they get displayed. But if we close the app and open it again, the data is gone. This is where in the next chapter, we will use CoreData to make our data persistent. We will also implement the 'Completion' of an item by marking 'To-Do's as 'Done'.

Summary

In this chapter, we have built the classic 'To Do List' app using tableviews. We implemented adding to-do items and reloading the table view with the newly added items. In the next chapter, we will address the data persistency problem by using Core Data.

Chapter 5: Persistent Data Using Core Data

In this chapter, we will be using Core Data to keep our data persistent, i.e. we will be able to save our to-do items and retrieve them even after we close and open our app. Currently when we close our app, all data is gone. Core Data allows us to take the objects we have created and saved them into an internal database. We later retrieve the objects from the database when we need to.

The important thing to note is that we should have 'checked' the 'Use Core Data' checkbox when we first created our project (fig. 1). We have done this in the previous chapter.

Product Name:	TodoList
Team:	None
Organization Identifier:	com.greglim.helloworld
Bundle Identifier:	com.greglim.helloworld.TodoList
Interface:	Storyboard
Life Cycle:	UIKit App Delegate
Language:	Swift
	☑ Use Core Data
	☐ Host in CloudKit
	☐ Include Tests

Figure 1

When we create a project with Core Data, we have a file called *<project name>.xcdatamodeld* or in our case, *ToDoList.xcdatamodeld* (fig. 2).

▼ ToDoList
　▼ ToDoList
　　AppDelegate.swift
　　SceneDelegate.swift
　　Main.storyboard
　　ToDoDetailsViewController.swift
　　AddToDoViewController.swift
　　ToDo.swift
　　ToDoTableViewController.swift
　　Assets.xcassets
　　LaunchScreen.storyboard
　　Info.plist
　　ToDoList.xcdatamodeld
　▶ Products

Figure 2

This is the file where we create 'entities' to be saved into Core Data. You can think of a Core Data 'entity' as the same as a 'Class'. Currently we have *ToDo.swift* to represent a to-do object, but now, we will have an entity to represent our to-do object.

Create an entity by clicking on the 'Add Entity' button 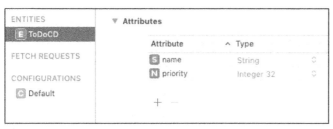 at the bottom of the *ToDoList.xcdatamodeld*. By default, 'Entity' will be created for you. Rename it to '*ToDoCD*' (CD abbreviation for CoreData). Next in 'Attributes', add the two attributes *name* and *priority* and specify their type to be *String* and *Integer 32* respectively (fig. 3).

Figure 3

In the following sections, we will be adding Core Data functionality to our *AddToDoViewController*, *ToDoTableViewController* and *ToDoDetailsViewController* where we can mark our to-dos items as done.

Saving into Core Data

Anytime we work with Core Data, we are working with a *ManagedObjectContext* which acts as a bridge from our app to the database. We illustrate this by first implementing Core Data for adding a to-do item. Fill in the following codes into *addTapped* of *AddToDoViewController.swift*:

```swift
@IBAction func addTapped(_ sender: Any) {

    if let context = (UIApplication.shared.delegate as?
AppDelegate)?.persistentContainer.viewContext
    {
        let newToDo = ToDoCD(context: context)
        newToDo.priority = Int32(prioritySegment.selectedSegmentIndex)

        if let name = nameTextField.text {
            newToDo.name = name
        }
        (UIApplication.shared.delegate as? AppDelegate)?.saveContext()
    }
    navigationController?.popViewController(animated: true)
```

```
    }
```

Code Explanation

To save data, we must have access to the core data stack. We do so via the *NSPersistentContainer* and *NSManagedObjectContext*. *NSPersistentContainer* contains the core data stack and *NSManagedObjectContext* acts like a doorway to allow users to save/fetch data from the core data stack.

AppDelegate.swift contains Core Data related methods and properties pre-defined when we first create our project. We will be using those methods and properties.

The code below is to show you how to access *NSManagedObjectContext* located in the *AppDelegate.swift* file

```
if let context = (UIApplication.shared.delegate as?
AppDelegate)?.persistentContainer.viewContext {
```

With the *context*, we then create a new *ToDo* entity object with *let newToDo = ToDoCD(context: context)*.

```
        let newToDo = ToDoCD(context: context)
```

That is, we no longer use *let newToDo = ToDo()*, in fact, we can delete *ToDo.swift* now.

```
        newToDo.priority = Int32(prioritySegment.selectedSegmentIndex)

        if let name = nameTextField.text {
            newToDo.name = name
        }
```

Like before, we assign the priority and name to the *newToDo* object. Take note however that we have to cast *prioritySegment.selectedSegmentIndex* to a *Int32* type because we had declared earlier in *ToDoList.xcdatamodeld* that *priority* is of type Integer 32 (fig. 4).

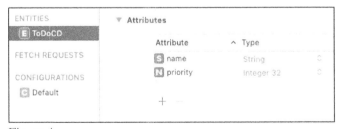

Figure 4

We then save the newly added *ToDo* item into Core Data with:

```
(UIApplication.shared.delegate as? AppDelegate)?.saveContext()
```

In the above code, we use the *AppDelegate.swift* built-in *saveContext* function to save data.

Note that we no longer have to do *toDoTableViewController?.toDos.append(newToDo)* since we save to our Core Data straight. We also don't call *reloadData (toDoTableViewController?.tableView.reloadData())* because we will be loading our tableview with data straight from Core Data later on.

```
navigationController?.popViewController(animated: true)
```

As per before, we pop the *Add* screen out of the stack to go back to the table view controller.

Pulling from Core Data

In this section, we will see how to pull data out of Core Data to display it in our table view controller.

In *ToDoTableViewController.swift*, previously we had *toDos* array containing *ToDo* objects. Change this to:

```
class ToDoTableViewController: UITableViewController {

    //var toDos: [ToDo] = []
    var toDoCDs: [ToDoCD] = []
```

Then, remove the following code in viewDidLoad():

```
override func viewDidLoad() {
    super.viewDidLoad()

    let toDo1 = ToDo()
    toDo1.name = "Write chapter one of iOS 14 book"
    toDo1.priority = 0
    let toDo2 = ToDo()
    toDo2.name = "Edit iOS 14 book"
    toDo2.priority = 1

    toDos = [toDo1,toDo2]    // fills up the array
```

}

Next, create a function *getToDos* with the following code:

```
func getToDos(){
    if let context = (UIApplication.shared.delegate as?
AppDelegate)?.persistentContainer.viewContext {
        if let toDosFromCoreData = try? context.fetch(ToDoCD.fetchRequest()){
            if let toDos = toDosFromCoreData as? [ToDoCD]{
                toDoCDs = toDos
                tableView.reloadData()
            }
        }
    }
}
```

Code Explanation

```
var toDoCDs = [ToDoCD]()
```

First, we change the *toDoCDs* array to contain *ToDo* Core Data objects instead of the *ToDo* object.

```
if let context = (UIApplication.shared.delegate as?
AppDelegate)?.persistentContainer.viewContext
```

As per before, whenever working with Core Data we need to have the *ManagedDataContext*.

```
if let toDosFromCoreData = try? context.fetch(ToDoCD.fetchRequest()){
```

We then attempt to fetch the request for *ToDo* Core Data objects. We add *try?* before the fetch to handle possible exceptions thrown.

```
            if let toDos = toDosFromCoreData as? [ToDoCD]{
                toDoCDs = toDos
                tableView.reloadData()
            }
```

We then assign the fetched *ToDo* Core Data objects to *toDoCDs* array following which we call *tableView.reloadData()* to reload the table view.

*Note: if Xcode shows an error on ToDoCD saying something like "it cannot find reference to ToDoCD object", just try cleaning the Build Folder and Build again. And if that doesn't work, you might need to exit and re-open Xcode.

viewWillAppear

Now, who will call *getToDos*? We want to get the to-dos each time the view appears. Thus, we implement the method in *viewWillAppear* which gets called whenever the view controller shows up and in it, we call *getToDos* as shown:

```
override func viewWillAppear(_ animated: Bool) {
    getToDos()
}
```

Changing from *ToDo* to *ToDo* Core Data

Changing our *ToDo* to *ToDo* Core Data would mean that we have to make small changes to *cellForRowAt*, *numberOfRowsInSection*, *didSelectRowAt* and *prepare*. So make the following changes:

numberOfRowsInSection

```
override func tableView(_ tableView: UITableView, numberOfRowsInSection
section: Int) -> Int {
    //return toDos.count
    return toDoCDs.count
}
```

cellForRowAt

The changes in *cellForRowAt* is mainly that *selectedToDo.name* now returns an optional, thus we have to use *if let name = selectedToDo.name{...}*. The rest remains the same as before:

```
override   func   tableView(_   tableView:   UITableView,   cellForRowAt
indexPath: IndexPath) -> UITableViewCell {
    let cell = UITableViewCell()
    //let selectedToDo = toDos[indexPath.row]
    let selectedToDo = toDoCDs[indexPath.row]

    if selectedToDo.priority == 1{
        if let name = selectedToDo.name{
            cell.textLabel?.text = " ! " + name
        }
```

90

```
        }
        else if selectedToDo.priority == 2{
            if let name = selectedToDo.name{
                cell.textLabel?.text = "!!" + name
            }
        }
        else{
            if let name = selectedToDo.name{
                cell.textLabel?.text = name
            }
        }
        return cell
    }
```

didSelectRowAt

```
    override func tableView(_ tableView: UITableView, didSelectRowAt
indexPath: IndexPath) {
        //let selectedToDo = toDos[indexPath.row]
        let selectedToDo = toDoCDs[indexPath.row]
        performSegue(withIdentifier: "moveToDetails", sender: selectedToDo)
    }
```

Marking our ToDo as Done (Deletion)

ToDoDetailsViewController.swift

```
class ToDoDetailsViewController: UIViewController {

    var toDoCD : ToDoCD? = nil
```

First in *ToDoDetailsViewController.swift*, we change the type of *toDo* variable to *toDoCD*. We also set it to an optional. This is because when we set a core data entity to be non-optional, we risk a crash when trying to save the context.

```
    override func viewDidLoad() {
        super.viewDidLoad()

        if let toDo = toDoCD {
            if toDo.priority == 1{
                if let name = toDo.name{
```

```
                    toDoLabel.text = " ! " + name
                }
            }
            else if toDo.priority == 2{
                if let name = toDo.name{
                    toDoLabel.text = "!!" + name
                }
            }
            else{
                if let name = toDo.name{
                    toDoLabel.text = name
                }
            }
        }
    }
```

In the above *viewDidLoad()*, because *toDoCD* is now an optional, we have to first unwrap it first with: *if let toDo = toDoCD {...}*. The code inside the curly braces remain largely the same.

doneTapped

We then implement the delete from Core Data in *doneTapped* as shown below:

```
    @IBAction func doneTapped(_ sender: Any) {
        if let context = (UIApplication.shared.delegate as?
AppDelegate)?.persistentContainer.viewContext {
            if let toDo = toDoCD{
                context.delete(toDo)
            }
            (UIApplication.shared.delegate as? AppDelegate)?.saveContext()
        }
        navigationController?.popViewController(animated: true)
    }
```

Code Explanation

```
    if let context = (UIApplication.shared.delegate as?
    AppDelegate)?.persistentContainer.viewContext
```

As always when working with Core Data, we retrieve the *ManagedDataContext.*

```
    if let toDo = toDoCD{
```

```
                context.delete(toDo)
            }
            (UIApplication.shared.delegate as? AppDelegate)?.saveContext()
        }
        navigationController?.popViewController(animated: true)
```

We then unwrap *toDoCD*, delete it from the context, save it and finally pop the detail view controller to reveal the table view controller.

prepare

Lastly, back in *ToDoTableViewController*, we need to make the below small change of changing the type to *ToDoCD*.

```
    override func prepare(for segue: UIStoryboardSegue, sender: Any?) {
        ...

        if let detailsToDoViewController = segue.destination as?
ToDoDetailsViewController{
            if let selectedToDo = sender as? ToDoCD{
                detailsToDoViewController.toDoCD = selectedToDo
            }
        }
    }
```

Running your App

Now when you run your app, try adding some to do items. Your items will now be stored persistently, You can also mark items done so that they are cleared off the list!

Summary

In this chapter, we learned how to use Core Data to keep our data persistent, how to save our to-do items and retrieve them even after we close and open our app. We covered how to work with a *ManagedObjectContext* to connect our app to the Core Data database. We explained how to pull data out of Core Data, display it in our table view controller and also how to delete items from Core Data. In the next chapter, we will revisit the To-Do List app, add images to each to-do item and implement swipe delete.

Chapter 6: To Do List with Images & Swipe Delete

In this chapter, we will improve on our To-Do list by letting users include images in their to-do items (fig. 1). When they add to-do items, they can also choose to include a related photo either through the camera or by picking an existing photo from their photo library.

Figure 1

And in the table view list of to-dos, the photos will be displayed along side the to-do items (fig. 2).

Figure 2

Deploying to a Device

To actually use the camera on our phone, we would have to deploy our app onto an actual phone. Alternatively, you can still run your app on the simulator but you will only be able to select photos from the photo library and not be able to use the camera function.

To begin deploying an app on an actual device, you need to have an Apple developer account. If you haven't got an account, sign up at developer.apple.com.

Using your developer account, login via Xcode by going to Xcode, *Preferences*, under 'Accounts', click on the '+' (fig. 3)

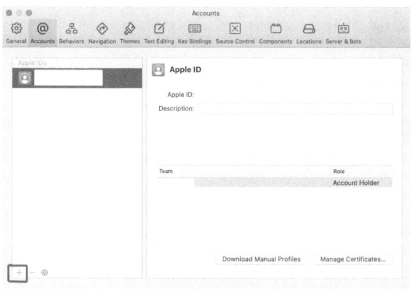

Figure 3

Select 'Apple ID' account (fig. 4) and sign in with your Apple ID.

Figure 4

Back in your project under 'Signing & Capabilities', 'Team', change it to your developer name (fig. 5).

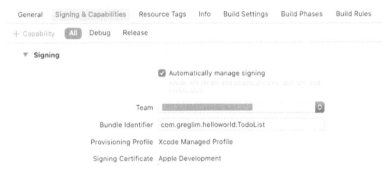

Figure 5

You should now be able to plug in your device via USB and select your device beside the buttons. Hit 'Run' and if all goes well, your app will run on your phone. If you encounter errors, follow the instructions as prompted by Xcode.

Adding an ImageView

In your existing To-do project from the last chapter, in the storyboard, drag an image view into the *AddToDoViewController*. Set the left/right constraints to be 50 and set the vertical spacing to the 'Add' button above (fig. 6).

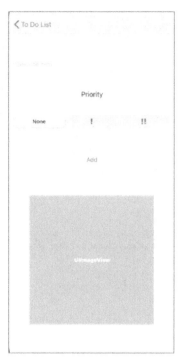

Figure 6

To make the image view into a nice square, specify the 'Aspect Ratio' between the height and width to be 1:1. To do so, select the image view, hold 'Ctrl' and drag click onto itself. There will be a popup, select 'Aspect Ratio' (fig. 7).

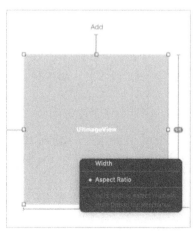

Figure 7

Under the 'Size Inspector', 📄 🕐 ⑦ 🖼 ⇄ ◣ ⓒ , 'Constraints', in 'Ratio', click on 'Edit' and specify 'Multiplier' to be 1:1 (fig. 8).

Figure 8

This way, our image view will be in a nice Instagram like square and the aspect ratio doesn't change across different device user interfaces.

Now, create an outlet for the image view in *AddToDoViewController.swift*:

```
@IBOutlet weak var imageView: UIImageView!
```

Camera and Organize Bar Buttons

Next, drag two bar button items onto the top of the screen (fig. 9).

Figure 9

To set the icon for the Bar button, select each of them and in 'Attributes Inspector', 'System Item', select 'Camera' for the right bar button and 'Organize' for the left bar button.

Core Data

To store images in Core Data, we have to go back to our *ToDoList.xcdatamodeld* to add the image attribute of Binary data type. So in 'Attributes', add the image attribute and specify its type to be 'Binary Data' (fig. 10). Other than image files like jpeg, the 'Binary Data' type can store other kinds of files like audio and video files.

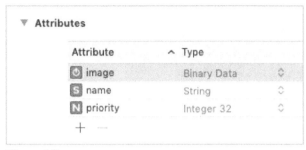

Figure 10

In the image attribute's 'Data Model Inspector', check the 'Allows Eternal Storage' checkbox (fig. 11). This option is useful when your file size gets very big and should be stored in external storage. The attribute instead just keeps a reference to the file in the database.

Figure 11

Selecting Photos with Image Picker Controller

Now, we will implement the action functions for both bar buttons. That is, when a user clicks on the *camera* bar button, the camera should appear and when the *organize* bar button is clicked, the photo library appears for the user to select the photo.

First in *AddToDoViewController.swift*, declare a *UIImagePickerController* object:

```
var pickerController = UIImagePickerController()
```

Next, create the actions for *camera* and *organize* bar buttons by drag clicking from them to *AddToDoViewController.swift*. Name the actions *cameraTapped* and *mediaFolderTapped*. Fill in the following code for the functions:

```
@IBAction func cameraTapped(_ sender: Any) {
    pickerController.sourceType = .camera
    present(pickerController,animated: true,completion: nil)
}
@IBAction func mediaFolderTapped(_ sender: Any) {
    pickerController.sourceType = .photoLibrary
    present(pickerController,animated: true,completion: nil)
}
```

Note that both functions are similar. The difference is that the *pickerController sourceType* for *cameraTapped* is *.camera* which opens up the camera and *sourceType* for *mediaFolderTapped* is *.photoLibrary* which opens up the photo library. The *present* method is the one which actually presents to the user. It accepts an argument *completion* which is an option to run some code after picking the photo. We won't be using this argument for now.

To display the selected photo in our image view, we need to set our current view controller as the delegate of the image picker controller. That is, the current view controller (the delegate) will be the one who will handle certain events raised by the image picker controller e.g. when someone takes a picture with the camera or chooses an existing photo from the photo library.

For the current view controller to be image picker delegate, we have to add the following to our class definition:

```
class AddToDoViewController:
UIViewController,UIImagePickerControllerDelegate,UINavigationControllerDe
legate
```

If you are asking, why do we need the *navigationcontroller* delegate? *UIImagePickerController* is a subclass of *UINavigationController*. It requires its delegate to implement the *UIImagePickerControllerDelegate* protocol while still implementing the *UINavigationControllerDelegate* protocol for its superclass.

And in *viewDidLoad()*, we need to set the current view controller to be the image picker controller delegate with:

```
override func viewDidLoad() {
    super.viewDidLoad()
    pickerController.delegate = self
}
```

We then need to implement the *didFinishPickingMediaWithInfo* function which is called after a photo is picked. When you begin to type in *didFinishPickingMediaWithInfo* into Xcode, it should autofill the function header and body. Implement *didFinishPickingMediaWithInfo* as follows:

```
func imagePickerController(_ picker: UIImagePickerController,
didFinishPickingMediaWithInfo info: [UIImagePickerController.InfoKey : Any]) {
    if let image = info[.originalImage] as? UIImage{
        imageView.image = image
    }
    pickerController.dismiss(animated: true, completion: nil)
}
```

imagePickerController delegate when called provides us with a variable *info* which contains various media meta data like *imageURL*, media type etc.

```
if let image = info[.originalImage] as? UIImage{
    imageView.image = image
}
```

We retrieve the original image with the key *originalImage* and unwrap it into *UIImage* type. We then assign the image to the image view.

```
pickerController.dismiss(animated: true, completion: nil)
```

After picking the image, we dismiss the image picker controller.

Running your App

When we run our app now, we will be able to select the photo we want and have it displayed on the image view (fig. 12).

102

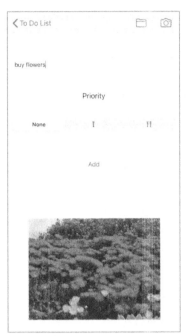

Figure 12

Saving our Image to Core Data

To save the selected image to our *ToDo* Core Data entity, add the following line:

```
@IBAction func addTapped(_ sender: Any) {
    if let context = (UIApplication.shared.delegate as?
    AppDelegate)?.persistentContainer.viewContext {
        let newToDo = ToDoCD(context: context)
        newToDo.priority = Int32(prioritySegment.selectedSegmentIndex)

        if let name = nameTextField.text {
            newToDo.name = name
        }
        newToDo.image = imageView.image?.jpegData(compressionQuality: 1.0)
        (UIApplication.shared.delegate as? AppDelegate)?.saveContext()
    }
    navigationController?.popViewController(animated: true)
}
```

UIImage has an instance method *jpegData(compressionQuality:)* which returns a data object containing the specified image in JPEG format. *jpegData* accepts an argument *compressionQuality* which is the quality of the resulting JPEG image, expressed as a value from 0.0 to 1.0. The value 0.0 represents the maximum compression (or lowest quality) while the value 1.0 represents the least compression (or best quality).

Displaying Images in the TableView

Now, we want to display each *to-do* image in the table view (fig. 13).

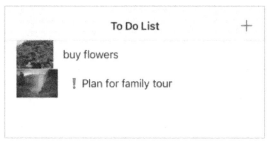

Figure 13

To do so, in the existing *cellForRowAt* code of *ToDoTableViewController.swift*, add:

```
    override func tableView(_ tableView: UITableView, cellForRowAt
indexPath: IndexPath) -> UITableViewCell {
        let cell = UITableViewCell()
        let selectedToDo = toDoCDs[indexPath.row]

        if selectedToDo.priority == 1{
            if let name = selectedToDo.name{
                cell.textLabel?.text = " ! " + name
            }
        }
        else if selectedToDo.priority == 2{
            if let name = selectedToDo.name{
                cell.textLabel?.text = "!!" + name
            }
        }
        else{
            if let name = selectedToDo.name{
                cell.textLabel?.text = name
            }
        }
```

104

```
    if let data = selectedToDo.image{
        cell.imageView?.image = UIImage(data:data)
    }
    return cell
}
```

The cell has an image view built into it which we can access via *cell.imageView*. If an image is set, it appears on the left side of the cell, before any label (fig. 14).

Figure 14

Swipe Delete

You might have seen apps that allow you to delete a row from a table view from swiping it to reveal a delete button (fig. 15).

Figure 15

To implement swipe delete in our app, we implement the *commit editingStyle* function. The function is actually available to us in the comments. We just have to uncomment it, and implement it with the following code:

```
override func tableView(_ tableView: UITableView, commit
    editingStyle: UITableViewCell.EditingStyle, forRowAt indexPath:
    IndexPath) {
```

```
if editingStyle == .delete {
    if let context = (UIApplication.shared.delegate as?
    AppDelegate)?.persistentContainer.viewContext {
        let selectedToDo = toDoCDs[indexPath.row]
        context.delete(selectedToDo)
        (UIApplication.shared.delegate as?
          AppDelegate)?.saveContext()
        getToDos()
    }
  }
}
```

The *commit editingStyle* function is called when someone wants to edit a certain row. The editing style can be delete or insert. In our case, we are just concerned with delete.

Thus check for a delete with *if editingStyle == .delete*. With this code in place, when you run your app now, you will reveal the delete button upon swipe. To actually delete the to-do from Core Data, we use the familiar code of:

```
if let context = (UIApplication.shared.delegate as?
AppDelegate)?.persistentContainer.viewContext {
    let selectedToDo = toDoCDs[indexPath.row]
    context.delete(selectedToDo)
    (UIApplication.shared.delegate as?
      AppDelegate)?.saveContext()
    getToDos()
}
```

We use *indexPath.row* to get the index of the selected row and retrieve the corresponding *ToDo* Core Data object from *toDoCDs* array. We then delete it from the context, save it, and call *getToDos* again to reload the data onto the table view.

Running your App

When you run your app now, you will be able to swipe delete a particular *todo* item.

Summary

In this chapter, we improved on our To-Do list from the last chapter by letting users include images in their to-do items. We also covered how to deploy our app to an actual device using a developer account. We used Core Data to store the images by using the binary data type attribute. We also covered how to implement swipe deleting a single table view row. In the next chapter, we will learn how to connect our app with an API to retrieve data.

Chapter 7: Connecting to an API: Cryptocurrency Price Tracker

In this chapter, we will learn how to connect our app with the Internet. We will be connecting to an API to get the price of different cryptocurrencies to display to the user. We will be using a single view controller that contains a picker with two components to select the cryptocurrency and in what real world currency we want the price of the cryptocurrency to be in.

In our app, we will let the user pick from a range of cryptocurrencies: BTC, ETH, XRP and BCH and real world currencies: "USD", "EUR", "JPY", "CHF" and the price retrieved will be displayed in a nicely formatted fashion (fig. 1).

Figure 1

Start a new 'App' project and name it 'Cryptocurrency Price Tracker'. As we won't be using Core Data, leave that check box unchecked.

Next, select the default view controller, go to 'Editor' > 'Embed In' and select 'Navigation Controller' just as what we have done in chapter two in our quotes app. A navigation bar will appear at the top of the view.

Proceed to drag and drop a description label on the top (text: "Choose Cryptocurrency & Currency"), a picker below and a price label below into the view like in figure 2.

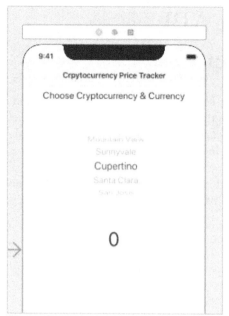

Figure 2

Set the left/right margins for all of them to be 10. For the description label on top, also set top margin to be 10. Set the vertical spacing for the picker and price label. Try out different device views to see how the controls display.

For the price label, increase the font size so that it is more easily seen as it is the main thing users are looking for.

Create the outlets for the currency picker and the price label.

```
class ViewController: UIViewController,UIPickerViewDelegate,
UIPickerViewDataSource {

    @IBOutlet weak var picker: UIPickerView!
    @IBOutlet weak var price: UILabel!
    ...
```

Selection Using Picker View

We will go through the steps of creating and populating our picker with the various currency options. The *UIPickerView* is a UI element used to make a selection from multiple choices (similar to a dropdown).

110

Let's first create the data that we are going to display in the picker. Fill in the following codes into *viewDidLoad*:

```
...
    var crpCcy: [String] = [];
    var ccy: [String] = [];
...

    override func viewDidLoad() {
        super.viewDidLoad()
        crpCcy = ["BTC", "ETH", "XRP", "BCH"]
        ccy = ["USD", "EUR", "JPY", "CHF"]
    }
```

Because we have two components in our picker (cryptocurrency and currency), we have declared two arrays each storing the range of values for each component.

Next, we add the code to connect the array's data to our picker. We first need to have our class conform to the *UIPickerViewDelegate* and *UIPickerViewDataSource* protocols. Add the following code to the class definition:

```
class ViewController: UIViewController,UIPickerViewDelegate,
UIPickerViewDataSource {
```

Similar to what we have done for *UITableViewDelegate* and *UITableViewDataSource*, what we are saying is that the *ViewController* class now conforms to the 'rules' of being a picker view delegate and datasource in that it will handle the events raised by the *UIPickerView* class. We then set the current view controller instance (*self*) as the delegate and datasource of the picker view with the following lines:

```
    override func viewDidLoad() {
        super.viewDidLoad()
        crpCcy = ["BTC", "ETH", "XRP", "BCH"]
        ccy = ["USD", "EUR", "JPY", "CHF"]

        self.picker.delegate = self
        self.picker.dataSource = self
    }
```

To handle the events raised by the *UIPickerView* class, we have to implement the following datasource methods:

111

```
func numberOfComponents(in pickerView: UIPickerView) -> Int {
    return 2
}

func pickerView(_ pickerView: UIPickerView, numberOfRowsInComponent
component: Int) -> Int {
    if (component == 0) {
        return crpCcy.count;
    }
    else {
        return ccy.count;
    }
}

func pickerView(_ pickerView: UIPickerView, titleForRow row: Int,
forComponent component: Int) -> String? {
    if (component == 0) {
        return crpCcy[row];
    }
    else {
        return ccy[row];
    }
}
```

numberOfComponentsInPickerView asks for the number of columns in our picker. In our case we return 2 meaning 2 columns. If you wanted a picker for date, you might have 3 components of year, month and day.

numberOfRowsInComponent asks for the number of rows of data in each component. So we return the array count. Because we have two components and each component could possibly have different number of rows, we use an if-else depending on *component* to return the correct number of rows. For e.g. for the first component *(component == 0)*, we return *crpCcy.count*.

```
if (component == 0) {
    return crpCcy.count;
}
else {
    return ccy.count;
}
```

If there were more components, we will need *component* == *1* and so on.

titleForRow asks for the title to be displayed in each row. This is where we retrieve the data from the array and return it. Similar to *numberOfRowsInComponent*, if component == 0, we return *crpCcy[row]* *(crpCcy = ["BTC", "ETH", "XRP", "BCH"])*, else *return ccy[row]* *(ccy = ["USD", "EUR", "JPY", "CHF"])*. This is to ensure that we retrieve from the appropriate array based on which component is selected.

```
if (component == 0) {
    return crpCcy[row];
}
else {
    return ccy[row];
}
```

If you run your app now, you will see the options populated in your picker. But how do we select what the user has picked?

didSelectRow

To detect what the user has selected in our picker, we have to implement the delegate method, *didSelectRow*:

```
func pickerView(_ pickerView: UIPickerView, didSelectRow row: Int,
inComponent component: Int) {
    getPrice(crpCcy: crpCcy[picker.selectedRow(inComponent: 0)],ccy:
        ccy[picker.selectedRow(inComponent: 1)])
}
```

In *didSelectRow* row, we get the selected row in the first component using *picker.selectedRow(inComponent: 0)* and retrieve the corresponding element from the *crpCcy* array. We do the same for the second component using *picker.selectedRow(inComponent: 1)* to retrieve from *ccy* array. With the selected cryptocurrency and currency, we then call a method *getPrice* which retrieves the price from our API.

Retrieving Data from an API

How do we get the cryptocurrency price? We use *https://min-api.cryptocompare.com/*, a free API for getting cryptocurrency live pricing data. There are APIs for all kinds of data, e.g. weather information, Twitter tweets, Facebook posts. An API could be seen as a way for computers to talk to computers.

The default sample API provided in the site is:
https://min-api.cryptocompare.com/data/price?fsym=BTC&tsyms=USD,JPY,EUR
which returns the result:

USD: 6381.51
JPY: 716836.61
EUR: 5621.46

That is, get the price of Bitcoin in USD, JPY and EUR. We can retrieve the price of Ethereum by specifying *fsym=ETH* instead of *fsym=BTC*:

https://min-api.cryptocompare.com/data/price?fsym=ETH&tsyms=USD,JPY,EUR

Now having this API, we next go on to implement our *getPrice* method.

getPrice: URLSession

Fill in the following codes into *getPrice*:

```
func getPrice(crpCcy: String, ccy: String){
    if let url = URL(string: "https://min-api.cryptocompare.com/data/price?fsym="
        + crpCcy + "&tsyms=" + ccy){
        URLSession.shared.dataTask(with: url) { (data, response, error) in
            if let data = data {
                print("connected")
            }
            else{
                print("wrong =(")
            }
        }.resume()
    }
}
```

Code Explanation

```
if let url = URL(string: "https://min-api.cryptocompare.com/data/price?fsym=" +
crpCcy + "&tsyms=" + ccy){
```

We create a *URL* object from the API url string. The url string takes in the user selected cryptocurrency and real world currency of choice and concatenates into the url string.

```
        URLSession.shared.dataTask(with: url) { (data, response, error) in
```

We then input the url into *URLSession.shared.dataTask* which creates the task for calling a web service endpoint on a remote server. *URLSession* is used to create the session. With the session, we then call *dataTask* to create a data task.

URLSession.shared.dataTask expects us to provide a completion handler which is a function that gets called when the request completes. When the request completes, we are provided the data retrieved, the response and any errors thrown.

```
URLSession.shared.dataTask(with: url) { (data, response,
error) in
    if let data = data {
        print("connected")
    }
    else{
        print("wrong =(")
    }
}.resume()
```

So in the call back function we provide, we check if data is not nil, in which case our request got connected successfully and we print "connected". Else, we print "wrong" which means there is something wrong with our url. Because tasks are created in a suspended state, we start the task with *resume()*.

If we run our app now, it should successfully connect and print "connected".

getPrice: JSON

Now how do we process the data retrieved? The API retrieves our data in a format called JSON. JSON is an easy way to represent data that facilitates data transfer between computers. It's a standard format which many APIs use. In Swift, we can easily access JSON data using dictionaries. That is, data is stored in key-value pairs. For example, the API call:

https://min-api.cryptocompare.com/data/price?fsym=BTC&tsyms=USD,JPY,EUR

returns us the data:

```
{
    "USD":6417.8,
    "JPY":723274.95,
    "EUR":5650.09
}
```

The data is returned in key-value pairs of ("USD":6417.8), ("JPY":723274.95) and ("EUR":5650.09). This fits perfectly with dictionaries in Swift which are collections whose elements are key-value pairs as well.

To use dictionaries to structure our returned JSON data, we do the following:

```
func getPrice(crpCcy: String, ccy: String){
    if let url = URL(string: "https://min-
api.cryptocompare.com/data/price?fsym=" + crpCcy + "&tsyms=" + ccy){
        URLSession.shared.dataTask(with: url) { (data, response, error) in
            if let data = data {
                if let json = try? JSONSerialization.jsonObject(with: data,
                options:[]) as? [String:Double]{
                    if let price = json[ccy] {
                        print(price)
                    }
                }
            }
            else{
                print("wrong =(")
            }
        }.resume()
    }
}
```

We provide the returned data into *JSONSerialization.jsonObject(with: data…)*, and because we have no options, we provide an empty array. Because our data is in the JSON format of e.g. "USD":6417.8, we specify *as? [String: Double]*.

```
if let price = json[ccy] {
    print(price)
}
```

Because *JSONSerialization.jsonObject* potentially throws exceptions, we use the *try?* keyword. We then unwrap the JSON object and finally provide the key to retrieve the price value and print it to the console. When you run your app now, it will retrieve and print the price to the console!

Running Code on the Main Thread

Now, instead of printing the price in the console, we should be displaying the price in the label with

```
if let price = json[ccy] {
    self.price.text = "\(price)"
}
```

116

But if we try to do this, we will get an error:

UILabel.text must be used from main thread only

Now this has happened because we are now running code on the callback function provided to *URLSession.shared.dataTask*:

```
URLSession.shared.dataTask(with: url) { (data, response, error) in

   ...
}.resume()
```

Our code now is running on a separate thread to the main thread. To refer to our user controls, we need to be running on the main thread. To run code on the main thread, we enclosed in *DispatchQueue.main.async* as shown:

```
func getPrice(crpCcy: String, ccy: String){
    if let url = URL(string: "https://min-
api.cryptocompare.com/data/price?fsym=" + crpCcy + "&tsyms=" + ccy){
        URLSession.shared.dataTask(with: url) { (data, response, error) in
            if let data = data {
                if let json = try? JSONSerialization.jsonObject(with: data,
                options:[]) as? [String:Double]{
                    DispatchQueue.main.async {
                        if let price = json[ccy] {
                            self.price.text = "\(price)"
                        }
                    }
                }
            }
            else{
                print("wrong =(")
            }
        }.resume()
    }
}
```

Whatever is enclosed in *DispatchQueue.main.async* is run on the main thread.

And so, if you run your app now, our prices finally get displayed! However, it would be better if we format our prices with the correct currency symbol, e.g. USD with $, JPY with yen. To do so, we use the *NumberFormatter*.

Formatting Currencies

We use the *NumberFormatter* to format our currencies with the following code:

```
if let price = json[ccy] {
    let formatter = NumberFormatter()
    formatter.currencyCode = ccy
    formatter.numberStyle = .currency
    let formattedPrice = formatter.string(from: NSNumber(value:price))
    self.price.text = formattedPrice
}
```

Using the *NumberFormatter* object, we provide the currency code (e.g. USD, JPY, EUR) and specify *.currency* numberStyle and finally use the string function to get our formatted price, i.e. €185.47 (fig. 3).

Figure 3

Try it Yourself

Currently, after we fetch our prices, it does not get updated. Create a 'refresh' button to gets the latest price.

Summary

In this chapter, we learned how to connect our app with an API to get the price of different cryptocurrency live prices. We populated a picker with the various currency options and processed retrieved JSON data from the API using dictionaries. We also covered using the *NumberFormatter* to format prices based on their currencies. In the next chapter, we will build an image classification app using machine learning!

Chapter 8: Machine Learning with Core ML

In this chapter, we will learn about Core ML and Create ML, Apple's library to make machine learning easy for us.

Machine Learning taps on the idea that if you train a machine to do something repeatedly, it will eventually begin to make appropriate choices on its own. For example, if we show a machine thousands of pictures of cats and dogs, we eventually train the machine to be able to look at a picture and identify if it's a cat or dog. The set of pictures which a machine is trained on is called a model. You can either create your own model or use an existing model created by others. In this chapter, we will first go through usage of a model provided by Apple called the ResNet model that identifies from an image if it is a tree, animal, kind of food and more. You can easily search for more Core ML models on GitHub. Later on in the chapter, we will teach you how to create your models with Create ML.

Machine learning has many other applications. For example:
- if a picture contains nudity and thus filter out such images
- in auto answering of email/sms. Gmail provides the function where we can choose to reply with provided answers upon looking at the context of the email, for e.g. 'Sounds good', 'Thanks', etc.
- classifying newspaper articles into categories e.g. business, technology, politics

All these help in either automation or saving humans a bit of time in making decisions that humans would normally be making. In this chapter, we would be making an app which uses machine learning to recognize and classify pictures.

Downloading the ResNet Model

To begin, go to https://developer.apple.com/machine-learning/ (fig. 1)

Figure 1

Under the section 'Models', scroll down to the *ResNet50* model and click on 'View Models and Code Sample'(fig. 2).

Resnet50
Image Classification

A Residual Neural Network that will classify the dominant object in a camera frame or image.

Figure 2

Download the 'Resnet50.mlmodel' file (fig. 3).

Figure 3

CoreML in Action

Start Xcode and create a new App project called 'ImageRecognizer'. Uncheck the 'Core Data' check box since we will not be using it.

In the project, we include the downloaded ResNet50 model into our project by dragging the downloaded *Resnet50.mlmodel* file into Xcode. In the popup, check the 'Copy items if needed' and 'Add to targets' checkbox (fig. 4).

Figure 4

Resnet50.mlmodel will then be added to your project where you can see more details upon selecting it, for example, *name*, *type*, *size*, *description* and *license* (fig. 5).

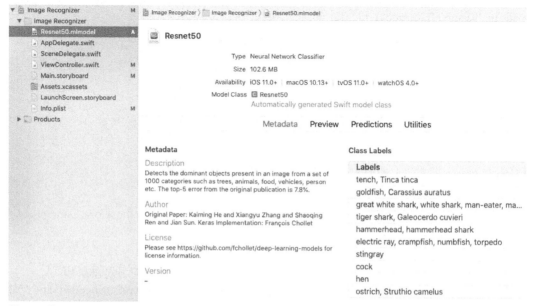

Figure 5

The more important part is the section 'Prediction' (fig. 6). In it contains the input/output of the model. For example, the model expects an image as input and will then output two fields, a *classLabel* of *String* type which is the most likely image category i.e. tree and also a *classLabelProbs Dictionary* which contains a collection of *String/Double* pairs which lists the image category with its probability i.e. Tree: 67.8%, Flower: 34.5% etc.

Metadata	Preview	Predictions	Utilities

Input

image
Image (Color 224 × 224)

Description
Input image of scene to be classified

Output

classLabelProbs
Dictionary (String → Double)

Description
Probability of each category

classLabel
String

Description
Most likely image category

Figure 6

ViewController.swift

In *ViewController.swift*, import *CoreML*, *Vision* and declare a *Resnet50* object with the following lines:

```
import UIKit
import CoreML
import Vision

class ViewController: UIViewController {
  var resnetModel = Resnet50()
```

Vision is the framework that works with Core ML to apply classification models to images, and to preprocess those images to make machine learning tasks easier and more reliable.

We next declare a method *classifyPicture* to process the image for classification:

```
func classifyPicture(image: UIImage){
   ...
}
```

We will fill the detail for *classifyPicture* later. *classifyPicture* will be called from *viewDidLoad*:

```
override func viewDidLoad() {
    super.viewDidLoad()

    if let image = imageView.image{
        classifyPicture(image:image)
    }
}
```

This of course presupposes that we have an image view in our *ViewController*. So proceed to add it in the storyboard and create the outlet for it in *ViewController.swift*:

```
import UIKit
import CoreML
import Vision

class ViewController: UIViewController {
```

```
var resnetModel = Resnet50()

@IBOutlet weak var imageView: UIImageView!

func classifyPicture(image: UIImage){

}

override func viewDidLoad() {
    super.viewDidLoad()
    if let image = imageView.image{
        classifyPicture(image:image)
    }
}
}
```

Using Image

We next implement filling up the image view with the phone camera or loading a picture from the library. The following steps ought to be familiar as we have previously done them when we added images into our to-do items back in chapter five.

We first need to declare an *UIImagePickerController* in *ViewController.swift* and extend *ViewController* with *UIImagePickerControllerDelegate, UINavigationControllerDelegate*:

```
class ViewController: UIViewController, UIImagePickerControllerDelegate,
UINavigationControllerDelegate {

    var resnetModel = Resnet50()
    var imagePicker = UIImagePickerController()
```

We also have to set *imagePicker.delegate* to *self* in *viewDidLoad*:

```
override func viewDidLoad() {
    super.viewDidLoad()

    imagePicker.delegate = self
    if let image = imageView.image{
        classifyPicture(image:image)
```

```
        }
    }
```

Linking Camera and Library Buttons

Next we create the action methods for the camera and photo library buttons which the user will use to select the image for classification. We will link the camera button to action method *cameraTapped* and the photo library button to action method *albumTapped*.

Before we create the action methods, we have to get permission from the user to access the photos. To do so, go to *Info.plist* and create two entries, 'Privacy – Camera Usage Description' and 'Privacy – Photo Library Usage Description' (fig. 7). In the 'Value' column, provide a description that is shown to the user when they request access.

▼ Information Property List		Dictionary	(16 items)
Privacy - Camera Usage Description	⌄	String	We need access to your camera to classify your photo
Privacy - Photo Library Usage Description	⌄	String	We need access to your photo library to classify an existing photo

Figure 7

Once that's done, proceed to create the action methods *albumTapped* and *cameraTapped* by first creating the buttons in the View Controller Storyboard. Drag a navigation bar into the view controller, and drag in two bar button item with the left bar button item's 'System Item' being 'Camera' and the right bar's being 'Organize' (fig. 8).

Figure 8

Proceed then to hold control and drag click from both buttons to create the action methods and fill them with the following code:

```
@IBAction func albumTapped(_ sender: Any) {
    imagePicker.sourceType = .photoLibrary
    present(imagePicker,animated: true,completion: nil)
}

@IBAction func photoTapped(_ sender: Any) {
    imagePicker.sourceType = .camera
    present(imagePicker,animated: true,completion: nil)
}
```

didFinshPickingMediaWithInfo

We next implement the *didFinshPickingMediaWithInfo* method which is called when we finish selecting our image.

Create the *didFinishPickingMediaWithInfo* method with the following code:

```
func imagePickerController(_ picker: UIImagePickerController,
didFinishPickingMediaWithInfo info: [UIImagePickerController.InfoKey :
Any]) {

        if let image = info[.originalImage] as? UIImage{
            imageView.image = image
            classifyPicture(image: image)
        }

        picker.dismiss(animated: true, completion: nil)
}
```

Next, we implement our *classifyPicture* method.

ClassifyPicture

Fill in the below code into *classifyPicture*:

```
func classifyPicture(image: UIImage){
    if let model = try? VNCoreMLModel(for: resnetModel.model){
        let request = VNCoreMLRequest(model:model){(request,error) in
            if let results = request.results as?
[VNClassificationObservation]{
                print(results)
            }
        }

        if let imageData  = image.jpegData(compressionQuality: 1.0){
            let handler =
VNImageRequestHandler(data:imageData,options:[:])
            try? handler.perform([request])
        }
    }
}
```

Code Explanation

```
if let model = try? VNCoreMLModel(for: resnetModel.model){
```

We first get a model reference with *VNCoreMLModel* and providing *resnetModel* as the argument. *VNCoreMLModel* is the container for our Core ML model used with *Vision* requests.

```
let request = VNCoreMLRequest(model:model){(request,error) in
    if let results = request.results as?
[VNClassificationObservation]{
        print(results)
    }
```

With the model reference, we create a *VNCoreMLRequest* image analysis request for identification processing through a completion handler:

```
{(request,error) in
 if let results = request.results as? [VNClassificationObservation]{
    print(results)
 }
```

The completion handler is called when we get back some information after the request has been completed. When do we actually execute the request? We execute it with the *perform* method of *VNImageRequestHandler*:

```
if let imageData  = image.jpegData(compressionQuality: 1.0){
  let handler = VNImageRequestHandler(data:imageData,options:[:])
  try? handler.perform([request])
}
```

VNClassificationObservation

Now I want to focus back on the results returned by the *VNCoreMLRequest* request. Notice in the console the print results of *print(results)*. The data returned by the request is an array results that contains an array of *VNClassificationObservation* objects. For example, when I run my app and choose the following photo for classification (fig. 9):

Figure 9 (to make your image fill the entire image view, under 'Attributes Inspector', 'View – Content Mode', set to 'Scale To Fill')

I get the below results printed into the console:

```
[<VNClassificationObservation: 0x600002b2dc80> BD9B032D-FCD6-45DC-A169-
0B6476AF67E1 requestRevision=1 confidence=0.743795 "cliff, drop, drop-off",
<VNClassificationObservation: 0x600002b37a80> B6B18D4C-075E-4296-985E-
7FB635985C7C requestRevision=1 confidence=0.102215 "valley, vale",
<VNClassificationObservation: 0x600002b2e8b0> A4147898-72DE-4CA2-AADE-
D1D1B004429A requestRevision=1 confidence=0.055269 "dam, dike, dyke"
...]
```

I show only the first three observations but essentially you can see that its saying:

0.744795 probability that it's a cliff, drop, drop-off (pretty accurate!)
0.102215 probability it's a valley
0.055269 probability it's a dam and so on.

The results returned are in descending order of probability. That is, the most probable observation is the first item in the array i.e. 74.5% probability that it's a cliff.

To present the information more appropriately, we can use a *for* loop to iterate through the results:

```
let request = VNCoreMLRequest(model:model){(request,error) in
  if let results = request.results as? [VNClassificationObservation]{
    for result in results{
      print(result.identifier + "," + String(result.confidence))
    }
  }
}
```

And this would return us the following:

cliff, drop, drop-off,0.74379456
valley, vale,0.10221455
dam, dike, dyke,0.055268552
geyser,0.036434744
volcano,0.03093923
…

We can of course illustrate the list of results in a table view. But for our app, we will just retrieve the first element in the array (the most probably result) and display its classification in our navigation bar header. To do so, make the following changes:

```
let request = VNCoreMLRequest(model:model){(request,error) in
  if let results = request.results as? [VNClassificationObservation]{
    let result = results[0]
    self.navBar.topItem?.title = result.identifier
  }
}
```

Remember to first create an outlet *navBar* for the Navigation Bar to access it via *self.*

Now when you run the app, you will get the first and most probable classification in the header (fig. 10).

Figure 10

In the above sections, we learned about Core ML to implement machine learning in our image classification app. Although we have just used the ResNet50 model, go on to try different models and explore the different Machine Learning apps that you can possibly create.

Chapter 9: Augmented Reality with ARKit

Augmented Reality is a way of putting virtual objects inside the real world making it look like the real and virtual are merged together. It is prevalent in many of the apps we use for example, Pokemon Go and Instagram Augmented Reality stickers (fig. 1).

Figure 1

In this chapter, we explore ARKit features in iOS13. Apple has made it easy for developers to create apps using Augmented Reality by providing new tools, modules, frameworks in ARKit. We will be building an app where we detect an image and render a 3D model on top of that image. Furthermore, when the image is moved, the app should be able to track the image so that the 3D model rendered will move along.

You can imagine the usage of such applications, for example, a furniture shop app where we can render a 3D model of a furniture in your house and move the furniture around to see how it fits in your house before buying the actual furniture.

To run ARKit, you will need to deploy your app to an iPhone 6S or above as ARKit requires the A9 chip or above hardware. If you don't yet have an ARKit enabled device, just follow along first and learn the concepts.

Adding Tracking Images

First, create a new Xcode project and choose *Augmented Reality App* (fig. 2).

Choose a template for your new project:
iOS
watchOS
tvOS
macOS
Cross-platform
Filter
Application
Single View App
Game
Augmented Reality App
Document Based App
Master-Detail App
Page-Based App
Tabbed App
Sticker Pack App
iMessage App
Framework & Library
Cocoa Touch Framework
Cocoa Touch Static Library
Metal Library
Cancel
Previous
Next

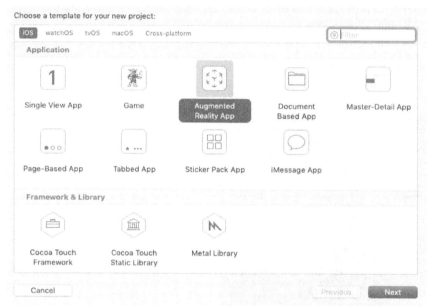

Figure 2

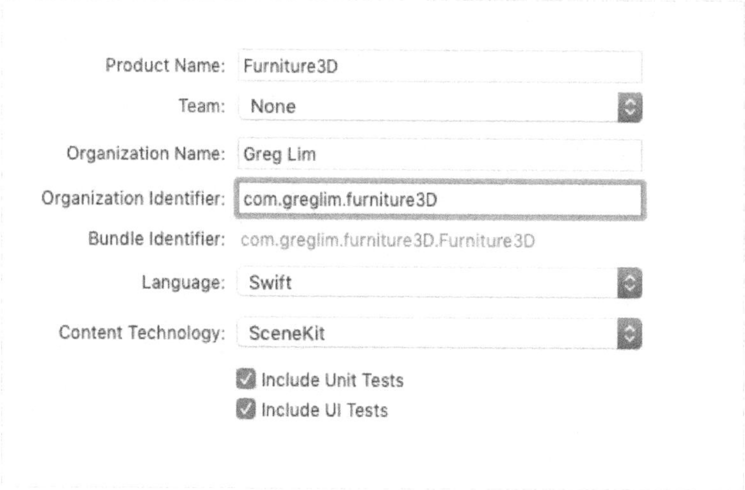

Product Name:
Furniture3D
Team:
None
Organization Name:
Greg Lim
Organization Identifier:
com.greglim.furniture3D
Bundle Identifier:
com.greglim.furniture3D.Furniture3D
Language:
Swift
Content Technology:
SceneKit
Include Unit Tests
Include UI Tests

Figure 3

Name the app 'Furniture3D' and in the 'Content Technology' field, select 'SceneKit' (fig. 3). There are three options provided in 'Content Technology': 'Metal', 'SpriteKit' and 'SceneKit'. *Metal* is a Graphical Processing Unit (GPU) accelerated graphics API that lets you control the GPU to push the hardware to its limits and control how your graphics would work. It is similar to OpenGL. *SpriteKit* and *SceneKit* are frameworks resting on top of *Metal*. Apple created template boilerplate code that has many of the basic implementation in game development to save developers much time and frustration to create games. *SpriteKit* was created for mainly 2D games and *SceneKit* is mainly for 3D games. In this chapter, we will focus on using *SceneKit* with ARKit.

Fill in your own 'Team'. Click 'Next'.

In the new project created for you in Xcode, you will see a *art.scnassets* folder. This is where you store your 3D assets. In it, there is a default 3D model *ship.scn* provided (fig. 4) and its texture map *texture.png*.

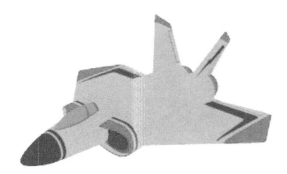

Figure 4

texture.png determines how the ship is going to look. If you select the model and look under 'Material Inspector', 'Properties, 'Diffuse' property, *texture.png* is specified there (fig. 5). The *diffuse* property describes the amount and color of light reflected equally in all directions from each point on the material's surface. It can be thought of as a material's "base" color or texture.

Figure 5

We will be adding our own 3D assets later. For now, we will use this model as a quick start. Next, we will add the image we want to detect. You can use any physical image or card that you have for the tracking image, for example, business cards, playing cards. It doesn't have to be specially made, just any ordinary card will do and that's the practicality of Apple's ARKit. For example, I have a picture of a fighter jet (fig. 6). Use a camera to take a picture of it getting it as big as you can and lining it up squarely.

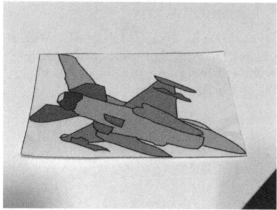

Figure 6

If you are taking the picture using your iPhone (which I assume you most probably are), your picture will be in a format called ".HEIC". We will have to convert this into a PNG format. To do so, simply open your image with the 'Preview' app in Mac.

In 'Preview', crop your photo to remove the surroundings so that what's left is just the picture itself and go to 'File', 'Export' and in 'Format' field, choose PNG and select 'Save' (fig. 7).

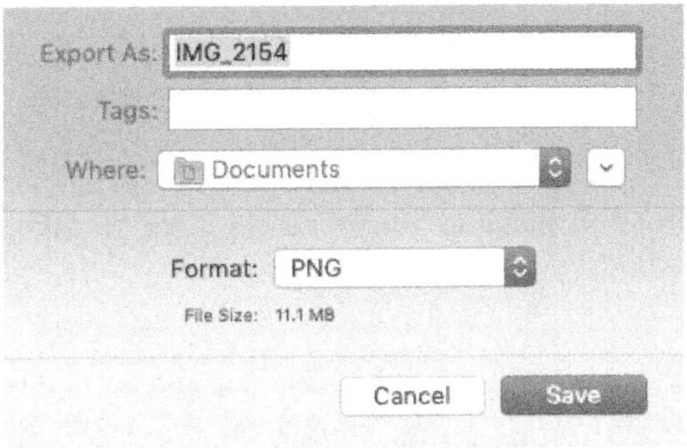

Figure 7

Your image will be exported to PNG in the specified folder.

Back in Xcode, select the *Assets.xcassets* folder, right click on the white section and select 'New AR Resource Group' (fig. 8).

Figure 8

A folder called 'AR Resource' will be created for you. You can change the folder name to anything you want. This folder is where you store tracking images. Drag and drop the card picture image file into this group.

Xcode might prompt you with a warning that you have not specified a physical width and height for the image (fig. 9).

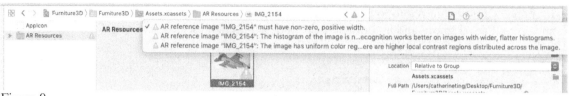

Figure 9

This is for ARKit to approximate how big the card should be to facilitate image recognition. So take a ruler, measure the card's dimensions and fill in the dimensions (fig. 10 note the units).

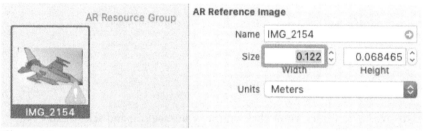

Figure 10

There might also be other warnings for example (fig. 9), the image recognition works better on images with wider, flatter histograms, the image has uniform colour contrasts etc. So make sure that you follow the guidance of these warnings. Else, the detection will not be so smooth.

ViewController.swift

Next, go to *ViewController.swift*. You will see some default code:

...

```
class ViewController: UIViewController, ARSCNViewDelegate {

    @IBOutlet var sceneView: ARSCNView!

    override func viewDidLoad() {
        super.viewDidLoad()

        // Set the view's delegate
        sceneView.delegate = self
```

138

```
        // Show statistics such as fps and timing information
        sceneView.showsStatistics = true

        // Create a new scene
        let scene = SCNScene(named: "art.scnassets/ship.scn")!

        // Set the scene to the view
        sceneView.scene = scene
    }

    override func viewWillAppear(_ animated: Bool) {
        super.viewWillAppear(animated)

        // Create a session configuration
        let configuration = ARWorldTrackingConfiguration()

        sceneView.session.run(configuration)
    }
```

...

Code Introduction

```
    @IBOutlet var sceneView: ARSCNView!
```

We have an outlet *sceneView* of *ARSCNView* type which links to a *SceneKit View* in the default View Controller in *main.storyboard* (fig. 11).

Figure 11

Apple has by default included an *ARKit* SceneKit View which you can search for in the Object Library using 'ARKit' (fig. 12). The *ARKit SceneKit View* and *SpriteKit View* are views to display your AR SceneKit or SpriteKit content respectively.

Figure 12

```
sceneView.delegate = self
```

We set *sceneView.delegate* to the current View Controller with *self*. Thus this controller conforms to *ARSCNViewDelegate* (*class ViewController: UIViewController, ARSCNViewDelegate*) and there's a whole set of delegate methods e.g. *didFailWithError*, *sessionWasInterrupted*, *sessionInterruptionEnded*. We will not be actually using these methods and can delete them if we want to.

```
sceneView.showsStatistics = true
```

sceneView.showsStatistics shows statistics such as fps and timing information while AR is rendered on your app. You can choose to remove this if you want.

```
// Create a new scene
let scene = SCNScene(named: "art.scnassets/ship.scn")!

// Set the scene to the view
sceneView.scene = scene
```

We create a new scene with *ship.scn* and open that scene in *sceneView*.

If you run your app on an ARKit enabled phone now, the ship 3D model will be rendered on to the camera (fig. 13).

Figure 13

Notice that the ship 3D model is fixed and cannot be moved around. For now, we don't want this default model showing. We remove it with the following codes:

...

```
class ViewController: UIViewController, ARSCNViewDelegate {
```

```
@IBOutlet var sceneView: ARSCNView!

override func viewDidLoad() {
    super.viewDidLoad()

    // Set the view's delegate
    sceneView.delegate = self

    // Show statistics such as fps and timing information
    sceneView.showsStatistics = true

    // Create a new scene
    let scene = SCNScene(named: "art.scnassets/ship.scn")!

    // Set the scene to the view
    sceneView.scene = scene
}
...
```

And in *viewWillAppear*, add the following codes:

```
override func viewWillAppear(_ animated: Bool) {
    super.viewWillAppear(animated)

    // Create a session configuration
    let configuration = ARWorldTrackingConfiguration()
    if let imageToTrack =
ARReferenceImage.referenceImages(inGroupNamed: "AR
        Resources", bundle: Bundle.main){
        configuration.detectionImages = imageToTrack
        configuration.maximumNumberOfTrackedImages = 1
    }
    sceneView.session.run(configuration)
}
```

Code Explanation

```
if let imageToTrack =
ARReferenceImage.referenceImages(inGroupNamed: "AR
    Resources", bundle: Bundle.main){
    configuration.detectionImages = imageToTrack
```

What we have done is tell our app that the reference image it should track is located in 'Assets.xcassets', 'AR Resources'. *Bundle.main* refers to the location of our current project file.

```
configuration.maximumNumberOfTrackedImages = 1
```

142

We specify in our configuration how many images we intend to track. Later on, we will increase this to two images where we track and render two different images and 3D models.

Adding Planes into our Images

In the previous section, we incorporated our tracking images into our Xcode project so that it knows what images it should be tracking. In this section, we will start recognizing those images and turning them into a plane that we will set our 3D models on.

We will be implementing the

```
func renderer(_ renderer: SCNSceneRenderer, nodeFor anchor: ARAnchor) -> SCNNode?
```

renderer delegate method which gets called whenever an image is rendered. In ARKit, the image detected is known as an *anchor*. The 3D model to be rendered is known as the SCN node or SceneKit node. *renderer* method should provide a 3D model to the image detected. You can see that *renderer* returns a *SCNNode* object to be rendered on to the image.

Firstly, a new 3D object or SCN node is created with:

```
func renderer(_ renderer: SCNSceneRenderer, nodeFor anchor: ARAnchor)
-> SCNNode? {
    let node = SCNNode()

    return node
}
```

Now add the following codes:

```
func renderer(_ renderer: SCNSceneRenderer, nodeFor anchor: ARAnchor)
-> SCNNode? {
    let node = SCNNode()

    if let imageAnchor = anchor as? ARImageAnchor{
        let plane = SCNPlane(
            width: imageAnchor.referenceImage.physicalSize.width,
            height: imageAnchor.referenceImage.physicalSize.height)

        let planeNode = SCNNode(geometry: plane)
        planeNode.eulerAngles.x = -.pi/2
        node.addChildNode(planeNode)
    }
```

```
        return node
    }
```

Code Explanation

```
    if let imageAnchor = anchor as? ARImageAnchor{
```

The *anchor* reference of type *ARImageAnchor* is provided for us. We unwrap it and in it, create a plane from the image:

```
        let plane = SCNPlane(
            width: imageAnchor.referenceImage.physicalSize.width,
            height: imageAnchor.referenceImage.physicalSize.height)
```

By specifying width and height with *imageAnchor.referenceImage.physicalSize width* and *height* attributes, we ensure that the created plane is the same physical size as the reference image.

```
        let planeNode = SCNNode(geometry: plane)
        planeNode.eulerAngles.x = -.pi/2
```

planeNode.eulerAngles.x = -.pi/2 is needed because if we run the app now, we see that the plane gets rendered on top of the image but it is rendered vertically when we want it to be parallel to our card (fig. 14).

Figure 14

Thus, we have to rotate our plane 90 degrees anti-clockwise around the x-axis of the SceneKit coordinate system (fig. 15). The *eulerAngles* method allow us to rotate the plane around a specified axis. In our case, we specify *eulerAnglues.x* to rotate around the x-axis of the SceneKit coordinate system.

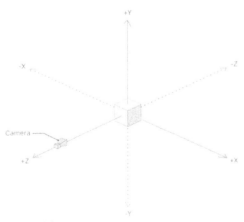

Figure 15

145

planeNode.eulerAngles.x expects the degree of rotation to be expressed in radians. To do so, we use *.pi* which represents 180 degrees. *-.pi/2* therefore equates to rotating it -90 degrees.

```
node.addChildNode(planeNode)
```

Finally, with the plane dimensions, we use it to create a plane node and add it to the root node. If you run the app now, you will see that the plane generated is attached and parallel to the card surface (fig. 16).

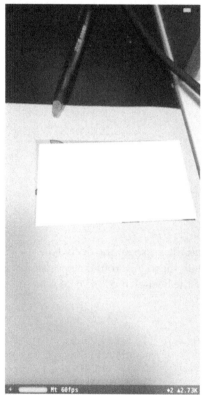

Figure 16

But because the plane is currently white and opaque, you can't see the card beneath anymore. Thus, we make the plane more transparent with the following line of code:

```
    func renderer(_ renderer: SCNSceneRenderer, nodeFor anchor: ARAnchor) ->
SCNNode? {
        let node = SCNNode()

        if let imageAnchor = anchor as? ARImageAnchor{
            let plane = SCNPlane(
```

146

```
                width: imageAnchor.referenceImage.physicalSize.width,
                height: imageAnchor.referenceImage.physicalSize.height)
        plane.firstMaterial?.diffuse.contents = UIColor(white: 1.0,alpha:
0.5)
            let planeNode = SCNNode(geometry: plane)
            planeNode.eulerAngles.x = -.pi/2
            node.addChildNode(planeNode)
        }

        return node
    }
```

If you run the app now, you can see it being more transparent as we have set *diffuse.contents* alpha transparency to 0.5 (fig. 17). If fully opaque, its alpha is 1. If fully transparent, its alpha is 0.

Figure 17

Now, we are ready to start rendering our 3D model on top of the card. We will use the plane to track, render and move our 3D model as the physical card moves along.

Rendering 3D model on top of card

As a quick start, we first render the default ship (looks more like a fighter jet!) model provided by Apple onto our card.

To do so, in *ViewController.swift*, add the following codes into *renderer*:

```
    func renderer(_ renderer: SCNSceneRenderer, nodeFor anchor: ARAnchor) ->
SCNNode? {
        let node = SCNNode()

        if let imageAnchor = anchor as? ARImageAnchor{
            let plane = SCNPlane(
                    width: imageAnchor.referenceImage.physicalSize.width,
                    height: imageAnchor.referenceImage.physicalSize.height)
            plane.firstMaterial?.diffuse.contents = UIColor(white: 1.0,alpha:
0.5)

            let planeNode = SCNNode(geometry: plane)
            planeNode.eulerAngles.x = -.pi/2
            node.addChildNode(planeNode)

            if let modelScene = SCNScene(named: "art.scnassets/ship.scn"){
                if let modelNode = modelScene.rootNode.childNodes.first{
                    planeNode.addChildNode(modelNode)
                    modelNode.eulerAngles.x = .pi/2
                }
            }
        }

        return node
    }
```

Code Explanation

```
        if let modelScene = SCNScene(named: "art.scnassets/ship.scn"){
```

We first create a *SCNScene* from our *ship.scn* file.

```
        if let modelNode = modelScene.rootNode.childNodes.first{
```

We then create a node *modelNode* to represent the ship model.

```
        planeNode.addChildNode(modelNode)
```

We then add the model node to the plane node.

```
        modelNode.eulerAngles.x = .pi/2
```

148

Again, the 3D model is rendered 90 degrees perpendicular to the plane. Thus we need to rotate it in the same way we did to rotate the plane.

Now if we run our app, we should have the plane rendered onto our card plane (fig. 18). And when we move our card, the model moves along as well!

Figure 18

You might experience that the detection might not be so smooth or that the rendered model sometimes disappears as you move the card around. This could be depending on factors such as the lighting in your environment (make sure your room is well lit), how defined the color contrasts are on the reference image etc. So make sure you follow whatever warning guidelines Xcode gives you.

In the next section, we will explore how to render a different model other than the default ship model.

Importing 3D models

In this section, we will see how to import different 3D models into our Xcode project. iOS uses the *.usdz* format for 3D objects. You can even send a 'usdz' file to an iPhone/iPad and straight away view it on the device without using any software. In this section, we will be getting a sample 3D model in *.usdz* format from https://developer.apple.com/arkit/gallery/ (fig. 19).

Figure 19

As the description on the ARKit website says, if you visit the page on your iPhone, you can see the 3D models rendered onto your phone screen where you can do pinch gestures to scale the model from small to large.

I will be downloading the red chair *.usdz* model (fig. 20). Download and drag it into the 'art.scnassets' folder in Xcode.

Figure 20

If you click it, Xcode will show the 3D model where you can move, drag it around to view it from different angles (fig. 21).

Figure 21

Next, we have to convert the *usdz* file into a *scn* file to use it in our Scene Kit framework used to render our object in ARKit. Select *redchair.usdz*, go to 'Editor', and select 'Convert to SceneKit scene file format (.scn)' and click 'Convert' (fig. 22).

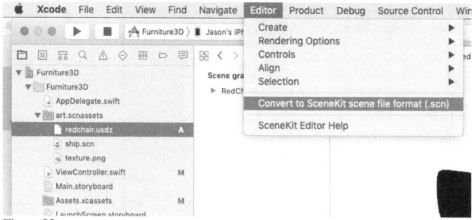

Figure 22

redchair.scn will now be available as a scene file like in the original sample *ship.scn*.

If you change your reference model name to,

```
if let modelScene = SCNScene(named: "art.scnassets/redchair.scn"){
```

and run your app, you should have the red chair model rendered instead. Now you will realise that the model rendered is too large so much so that you might not be able to see it! In this case, we need to scale it down.

To scale it down, select *redchair.scn*, and in the 'Scene Graph', go down to the deepest sub nodes and select all the sub folders, i.e. *RedChairFeet, RedChairLeg1, RedChairLeg2*. Go to 'Node Inspector' and under 'Scale', change the x, y and z values to: 0.05 (fig. 23). Do the same for the parent folder *RedChair* as well.

Figure 23

This will change the x,y and z values for all the sub models that make up the entire model so that the entire model is scaled down. And when you run your app now, you will see your red chair scaled down and rendered (fig. 24)!

Figure 24

How did I arrive at the scale value of 0.05? It is actually through trial and error. Other models might need a scale of different value. So just try different scale values and see if the model size turns out according to your expectation.

Rendering Multiple Models

In this section, we will have our app simultaneously detect two images to render two different 3D models.

We will be adding a television set to our project. First, find a card for image detection. I will be using one of my child's Chinese picture cards (fig. 25):

Figure 25

Add it to the *Assets.xcassets* folder and specify its physical width and height. Cards need not just be photos taken from a camera or scanner, they can also be image digital files added straight to Xcode. Realize that these are similar steps that we have done before to add a reference image for detection.

We will use the retro television model from the Apple AR Gallery (fig. 26). So go ahead and download its *usdz* file.

Figure 26

Drag the *usdz* model file into *art.scnassets* folder and convert it to the Scene Kit (scn) format (fig. 27) as what we have done earlier.

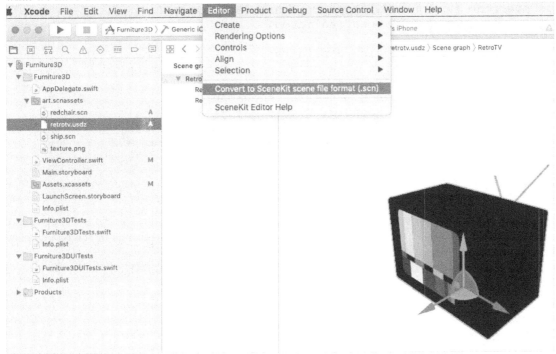

Figure 27

As the tv model is too big, scale the model size down to 0.05 for the x,y and z axis as we have done before (if 0.05 is still too big, you might have to try a scale factor of 0.01 instead - experiment around with the scale factor).

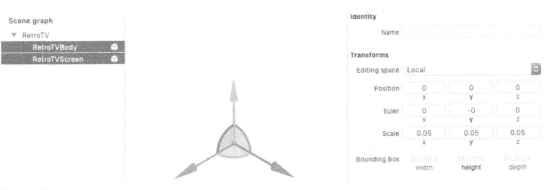

Figure 28

The first code change we make in *ViewController.swift* is to specify maximum number of tracked images = 2:

```
override func viewWillAppear(_ animated: Bool) {
    super.viewWillAppear(animated)

    // Create a session configuration
    let configuration = ARWorldTrackingConfiguration()
    if let imageToTrack =
ARReferenceImage.referenceImages(inGroupNamed: "AR
        Resources", bundle: Bundle.main){
        configuration.detectionImages = imageToTrack
        configuration.maximumNumberOfTrackedImages = 2
    }
    sceneView.session.run(configuration)
}
```

Next, to support multiple detection images and model renderings, we modify our *renderer* code such that when it detects a specific image, render a specific model. Thus, we need to name our reference images.

I have named my image *tvtable* and also added another image called *chair* for which I will render the red chair 3D model (fig. 29).

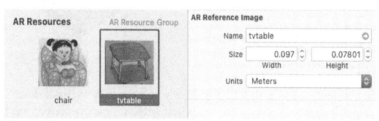

Figure 29

In *renderer*, use an *if-else* to check for the reference image name and render the appropriate model as shown below:

```
func renderer(_ renderer: SCNSceneRenderer, nodeFor anchor: ARAnchor) ->
SCNNode? {
    let node = SCNNode()

    if let imageAnchor = anchor as? ARImageAnchor{
        let plane = SCNPlane(
            width: imageAnchor.referenceImage.physicalSize.width,
            height: imageAnchor.referenceImage.physicalSize.height)
        plane.firstMaterial?.diffuse.contents = UIColor(white: 1.0,alpha:
0.5)

        let planeNode = SCNNode(geometry: plane)
        planeNode.eulerAngles.x = -.pi/2
```

156

```
            node.addChildNode(planeNode)

        if imageAnchor.referenceImage.name == "chair"{
            if let modelScene = SCNScene(named:
"art.scnassets/redchair.scn"){
                if let modelNode = modelScene.rootNode.childNodes.first{
                    planeNode.addChildNode(modelNode)
                    modelNode.eulerAngles.x = -.pi/2
                }
            }
        }
        if imageAnchor.referenceImage.name == "tvtable"{
            if let modelScene = SCNScene(named: "art.scnassets/retrotv.scn"){
                if let modelNode = modelScene.rootNode.childNodes.first{
                    planeNode.addChildNode(modelNode)
                    modelNode.eulerAngles.x = -.pi/2
                }
            }
        }
    }

    return node
}
```

And if you run your app now, you should have two models rendered together when you place the two cards together (fig. 30).

Figure 30

157

As you move both cards around, notice how the 3D models move along as well! And that is how you detect and render multiple models. Imagine the applications of such AR apps. For example, a furniture app that renders multiple furniture and allowing you to move the furniture around with the reference images, or AR games where you render character models to fight with one another!

Using Various 3D Model Formats

So far we have been using *usdz* model files provided from Apple's AR gallery. But if you take models from other sites, there are various other models formats like *.obj* or *.dae*. A good site to source for 3D models which other people have created is *www.turbosquid.com*. They have a huge catalog of 3D models both free and paid. Much of the models on *TurboSquid* are created using *Blender*, a great open source software that allows you to create lifelike 3D models. You can even create your own models using *Blender*. This of course won't be in the scope of this book.

To use the 3D models in Xcode, you need to first convert it into *.usdz* format. Apple has provided an Xcode command line tool to aid in this conversion.

First, in *Xcode*, *Preferences*, *Locations*, under 'Command Line Tools', select the latest Xcode version you have. It should be at least *Xcode 10* (fig. 31).

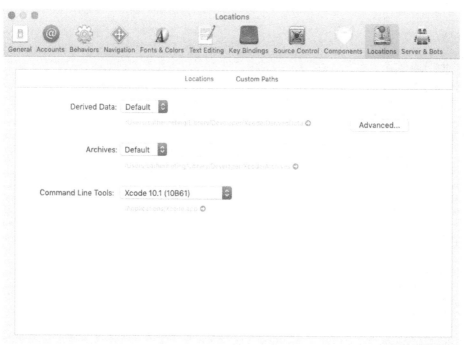

Figure 31

Now open Terminal and in it, run the 'xcrun usdz_converter' command to convert a model file from its source format to *usdz* format. The command goes like:

```
xcrun usdz_converter <.obj file> <usdz export location>
```

For example:

```
xcrun usdz_converter ~/Desktop/myFile.obj ~/Desktop/myFile.usdz
```

The *usdz* file will be created in your specified export location where you can proceed to add in Xcode.

Summary

In this chapter, we explored ARKit features in iOS13/14. We built an app where we detect an image and render a 3D model on top. We learned how to add tracking images, adding planes on those images and render 3D models on top of those planes. We covered importing 3D models into our project and simultaneously rendering multiple models on multiple tracking images. In the final chapter, we will learn how to publish our app onto the App Store.

Chapter 10: Publishing Our App on AppStore

To begin publishing our app on the Appstore, we have to enroll for the Apple developer program (https://developer.apple.com/programs/) which costs $99 per year. You will be granted an Apple developer account and given access to iTunes Connect (fig. 1 *itunesconnect.apple.com*).

Figure 1

In iTunes Connect, you can upload new apps, manage/update existing ones, check app analytics (how many people are using your app, the countries they are from), check sales/trends (how much app downloads, how much money you are making per day/week) and more. The other functions in iTunes Connect are quite self explanatory. You will also be asked to accept some terms and conditions, fill up tax and bank info etc.

When you are ready to upload your app, go to 'My Apps' and click on '+' on the top left. From the pop-up menu, choose 'New App'. In the form that appears (fig. 2), fill in the fields: app name. (app name has to be unique on the App store), *language* and other fields.

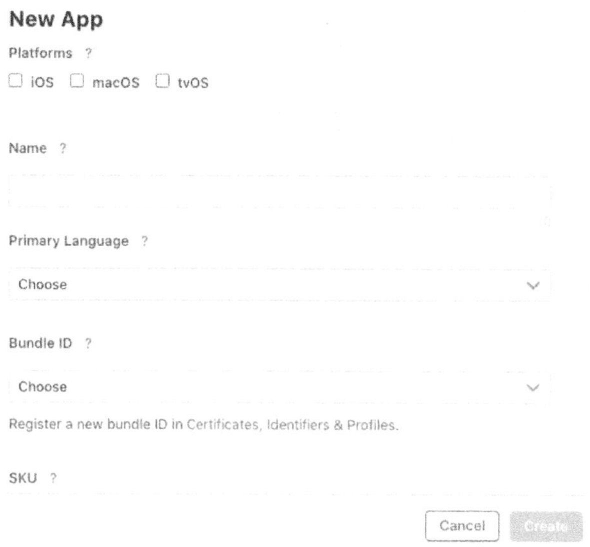

New App

Platforms   ?

☐ iOS   ☐ macOS   ☐ tvOS

Name   ?

Primary Language   ?

| Choose | ⌄ |

Bundle ID   ?

| Choose | ⌄ |

Register a new bundle ID in Certificates, Identifiers & Profiles.

SKU   ?

Cancel   Create

Figure 2

An important field to note is the 'Bundle Id' which is part of your app in Xcode, under 'Identity', 'Bundle Identifier'. If the Bundle Id doesn't show up automatically, you might have to follow the link to register a new Bundle Id and add it manually.

The 'SKU' (Stock Keeping Unit) field is for your reference only. It is a way of giving each of your app a unique ID amongst your other apps so you can identify it. When you have the form filled, click 'Create'. Where you can further choose the 'Category' your apps belong in and set the pricing of the app. You also need to provide screenshots for your app. Other fields to fill in are 'Description' and 'Keyword'.

When you are done, click on 'Submit Build'. 'Build' should show up once we submit using Xcode or Application Loader. In the next section, we show how to submit our app using Xcode.

Submitting our App using Xcode

Firstly, your app needs to be able to run properly and have no errors. Secondly, you should have your *Signing* and *Team* information setup in Xcode (fig. 3).

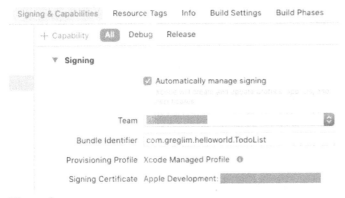

Figure 3

Also, make sure that you have got the correct Apple developer account signed in Xcode > *Preferences* > *Accounts* (fig. 4).

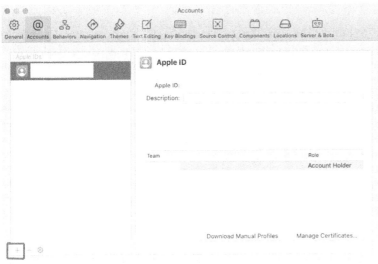

Figure 4

Since Xcode 11, publishing in Xcode is much simpler than in previous versions due to Xcode now being able to create certificates, profiles by itself. To create the full app archive, select 'Any iOS Device' from the list of simulators (fig. 5).

Figure 5

Then, go to 'Product' > 'Archive' to bundle the app. This process should take some time depending on the size of the app, but once it's done, the 'Archive' window will open, and this is where you can manage all the archives generated in Xcode. Select the archive and choose 'Distribute App'. Select 'iOS App Store' as the method of distribution.

Checks and validations will then occur to make sure that our app is valid for upload. If the validation fails, for example error in the code or missing assets like icons, go back and fix them. If the validation passes, the upload process will begin.

This will start a relatively long process of uploading the app to Apple's servers. It could take from ten minutes to an hour depending on the size of the project.

Once the app is uploaded, after five to ten minutes, the app should show up in 'All Builds' (fig. 6 back in iTunes Connect).

Figure 6

You can then select the *Save* button on the top right and the 'Submit for Review' button should turn blue. You will have a couple of 'Exported Compliance' and 'Advertising Identifier' related questions to answer about your app.

When that's done, you app will be on the waiting list for Apple's app review process. The review process took one to two weeks in the past but it has been improving in recent years. Once your app is approved, you will get an email notification.

Finally, 'App Review' info is where you put your contact information in so if your app review does not pass, Apple can contact you and tell you what to fix. Apple is very strict with their review guidelines and you should definitely read them (developer.apple.com/app-store/review/guidelines/).

Chapter 11: SwiftUI

SwiftUI was the biggest thing announced in WWDC 2019 in iOS13. It is a new and simple way to build user interfaces whether iOS, MacOS, TvOS or WatchOS (as we will show in the chapter regarding *Project Catalyst*) with a declarative Swift syntax that's easy to read and write. Previously, we relied on UIKit to design user interfaces but going forward, SwiftUI is the new visual way to make apps.

Like when Swift first came out, it wasn't perfect. If you wanted in-depth functionality, you would still have to use Objective-C. But as time progressed, Swift kept getting better and now it's the default language for iOS app development.

We're going to see a similar thing with SwiftUI being the new way to create apps user interfaces. Though it looks impressive (as we are about to find out), being in its beginning stages, it has teething problems. Yet, it is going to be the way forward to making apps and you should start learning that now. Going forward, we'll probably see lots of changes but it's only going to get better from here.

What makes SwiftUI so attractive is the functionality of *previews* that allow you to see visual updates to your code live. That is, when you make a code change, you don't have to hit the *run* button, wait for your simulator to show up and reflect the change. You make a code change and see exactly what it's going to look like straight in the preview. And that is so helpful as we are going to find out.

Requirements

The preview function is only available on the Mac OS Catalina operating system, i.e. version 10.15 onwards. You can still use previous Mac OS versions to run Xcode 12 and develop in SwiftUI. But you just won't see the live preview. And that's the reason we recommend having at least Catalina. With SwiftUI, it's much more fun if you have at least Catalina. In this chapter, I will be using Mac OS version 11, Big Sur.

Our First SwiftUI App

We're going to be creating a fun app that shows a list of favorite Pokemons. If you tap on a specific pokemon, it will show its picture with their type logo (e.g. fire type, water type) and some stats about them (fig. 1).

Figure 1

The objective of building this app is to introduce the capabilities of SwiftUI and realize how fun and quick this can happen. Now let's go ahead and make that app.

First, create a new Xcode project and choose 'App'. Make sure you select 'SwiftUI' under 'Interface' (fig. 2).

Figure 2

Notice that when we do so, we can only select 'Swift' in 'Language' (fig. 3). This is because SwiftUI is built natively with Swift.

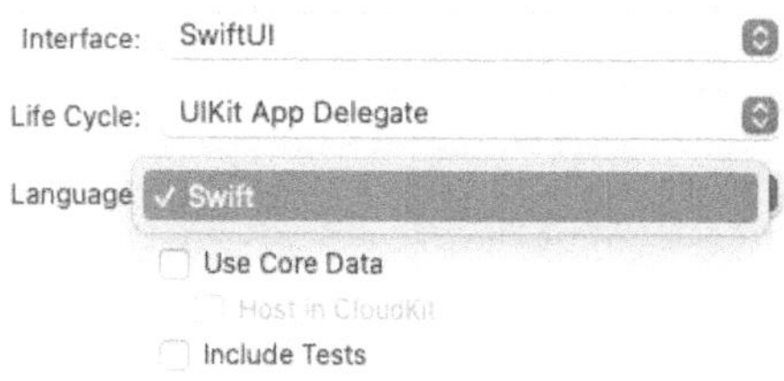

Figure 3

Leave the rest of the check boxes unchecked.

Once you have selected *Swift* as language and *SwiftUI* for Interface, give your project a name i.e. "PokemonSwiftUI" and click *Next*.

You'll notice a couple of things that are different here in Xcode. If you select *ContentView.swift*, you notice a preview of your app (fig. 4). If you don't see the preview, hit *resume* and it will create the preview. Note that it might take a while when you are opening the preview for the first time.

Figure 4

ContentView is the first view we're going to be working with. The preview allows us to check our UI without running the simulator. The preview uses whatever phone we have selected (fig. 5). So you could choose different devices including iPads.

PokemonSwiftUI ⟩ iPhone 11

Figure 5

Notice that whatever text we have in the below code is showing up in the preview. If I change the text from "Hello World" to "Hello SwiftUI", the preview text changes instantly (fig. 6).

```
struct ContentView: View {
    var body: some View {
        Text("Hello SwiftUI!")
            .padding()
    }
}
```

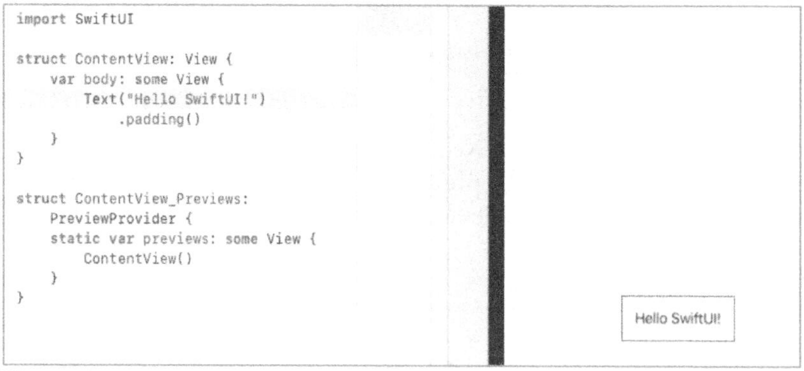

```
import SwiftUI

struct ContentView: View {
    var body: some View {
        Text("Hello SwiftUI!")
            .padding()
    }
}

struct ContentView_Previews:
    PreviewProvider {
    static var previews: some View {
        ContentView()
    }
}
```

Hello SwiftUI!

Figure 6

The real benefit of SwiftUI is making code changes to your project and seeing the results quickly reflected in the preview. Now let's do a little code explanation.

Code Explanation

```
struct ContentView: View {
    var body: some View {
        Text("Hello SwiftUI!")
            .padding()
    }
}
```

There's a *struct* called *ContentView*. There's nothing special about the name *ContentView*, we could rename this if we wanted to. The *ContentView strut* or structure conforms to the *View* protocol. Notice in the *struct* that there is a property called *body* of type *some View* and then there's curly brackets.

Before we go on, you might ask, we normally use *classes* for *views*. So why does SwiftUI use *structs* here? Firstly, structs are simpler and faster than classes. In SwiftUI, all views are trivial structs and take little resources to create. You avoid the overhead of inheriting from parent or grand-parent classes etc.

Secondly, and related to the first point, Swift by choosing to use structs encourages a more functional design approach where *views* become simpler and state data management is more organized as compared to having *classes* hold/change their own data freely which can lead to messier code. This however is out of the scope of this chapter. (In a future book that covers SwiftUI development in detail, I will cover state management in SwiftUI. Contact me at support@i-ducate.com to be notified for future releases.)

Moving back to the code within the curly brackets, it is returning a view for the property *body*. That is, implicit before *Text* is the *return* keyword. The code below is exactly the same as what we have seen earlier. But we do not have to specify *return* explicitly.

```
struct ContentView: View {
    var body: some View {
        return Text("Hello SwiftUI!")
            .padding()
    }
}
```

Text is essentially the new label. It just displays string text and can't accept any sort of user input. Finally, we use padding() to give the *Text* a padded spacing. Now, let's make some changes to show an image.

First, download an image and drag it into the white space area of 'Assets.xcassets' (fig. 7).

Figure 7

I have dragged in a charmander image file but you can download and drag your own pokemon image file in.

Now, delete *Text* and replace it with *Image* as shown below:

```
struct ContentView: View {
    var body: some View {
        Image("charmander")
            .padding()
    }
}
```

Provide the image name. When you do so, your picture should be displayed instantly (fig. 8). With SwiftUI, its simple to bring images in by just entering in the image name.

Figure 8

Your image might be taking up the entire screen as mine is. This is obviously what we do not want. To resize it, we specify that this image is resizable with:

```
Image("charmander").resizable().padding()
```

Unfortunately, it becomes stretched now (fig. 9). This is because by default, *resizable* allows the image to expand to fill all available space.

Figure 9

Rather, we want to adjust the frame. If you select the image on the preview, under the Attributes Inspector, you can see the different attributes you can change to alter the image. Under *Frame, Height,* specify *350* (fig. 10).

Figure 10

When you do so, Xcode generates the relevant code for you:

```
Image("charmander").resizable().padding().frame(height: 350.0)
```

And if you edit the value in the code, it will be reflected in the attributes inspector! This is extremely useful when you do not know what attributes are available for you in code but you can see them presented in the Attributes Inspector and have it generate code for you.

In SwiftUI, the code is the source of truth which affects the preview. Back in UIKit storyboards, sometimes there is some code behind the scenes in the view controller that you can't edit. For example, if you forget to delete a constraint, outlet or action, you might get some error and things can be quite messy. But for SwiftUI, the code is the single source of truth thus leading to lesser errors and more straight forward UI development.

Stacks

Next, we want to bring our image up to the top of the page and also add the 'type' picture (e.g. fire type, ice type) and name of the pokemon (fig. 11).

Figure 11

Before we go on, search and download a 'fire type' image (fig. 12) and drag it into 'Assets.xcassets' as what we have done before. I have named my image 'firetype'.

Figure 12

Now if we try to add the below code:

```
struct ContentView: View {
    var body: some View {
        Image("charmander").resizable().padding().frame(height: 350.0)
        Image("firetype").resizable().padding().frame(height: 350.0)
    }
}
```

You'll notice we're getting two views instead (fig. 13).

174

Figure 13

But what we want is the two images to be in a single view. What we should be doing is put multiple things inside a stack. There are the vertical, horizontal and Z stacks. For our app, we want to create a vertical stack because our images are stacked vertically.

Thus, enclose the two *Image* lines in a *VStack* as shown:

```
struct ContentView: View {
    var body: some View {
        VStack{
            Image("charmander").resizable().padding().frame(height: 350.0)
            Image("firetype").resizable().padding().frame(height: 350.0)
        }
    }
}
```

With this, the fire type image pops up just below the Charmander image (fig. 14).

175

Figure 14

Remember that we should pass back a single view. But a view can have a stack containing multiple views inside of it.

Next, we will add the pokemon name text into the app, inside of this vertical stack. Add the following:

```
struct ContentView: View {
    var body: some View {
        VStack{
            Image("charmander").resizable().padding().frame(height: 350.0)
            Image("firetype").resizable().padding().frame(height: 350.0)
            Text("Charmander")
        }
    }
}
```

Now, notice that everything's vertically centered in the middle (fig. 15). Stacks by default centers its content.

Figure 15

To push all of the visual controls to the top, we use something called a *Spacer*. Add it as shown below:

```
struct ContentView: View {
    var body: some View {
        VStack{
            Image("charmander").resizable().padding().frame(height: 350.0)
            Image("firetype").resizable().padding().frame(height: 350.0)
            Text("Charmander")
            Spacer()
        }
    }
}
```

Spacer() pushes everything up towards the top (fig. 16).

Figure 16

If we had *Spacer()* at the very top, i.e.:

```
VStack{
    Spacer()
    Image("charmander").resizable().padding().frame(height: 350.0)
    Image("firetype").resizable().padding().frame(height: 350.0)
    Text("Charmander")
}
```

everything gets pushed down to the bottom (fig. 17).

Figure 17

178

And if we were to put *Spacer()* between the two images, you would see that it would split the two of them (fig. 18).

Figure 18

Spacers essentially take up as much space as they possibly can. So, by putting this at the bottom, it's trying to fill the bottom with as much space as you can and thus push the rest of the controls to the top.

What we want to do next is to have the fire type image to be on top of the pokemon and in the background. So swop the two images:

```
VStack{
    Image("firetype").resizable().padding().frame(height: 350.0)
    Image("charmander").resizable().padding().frame(height: 350.0)
    Text("Charmander")
    Spacer()
}
```

Notice that at the top, there is some safe area the image avoids taking up (fig. 19).

Figure 19

To have the image fill up the top safe area, add the following:

```
VStack{
    Image("firetype").resizable().padding().frame(height: 400.0)
    Image("charmander").resizable().padding().frame(height: 350.0)
    Text("Charmander")
    Spacer()
}.edgesIgnoringSafeArea(.top)
```

We have found that a value of 400 for the *firetype height* makes the image look better. But go ahead and experiment different values for yourself to see which is most appropriate.

In this section, you've learned how to add multiple views inside a stack, how to use spacers and how to fill up the safe area space at the top. Next, we will see how to customize our views.

Customization

In this section, we want to see how we can customize views, i.e. have the picture of our Charmander in a circle, have a nice border and a shadow behind that (fig. 20).

Figure 20

We'd also like to have the font size of our pokemon's name be a little bigger. To increase the font and bold our pokemon name, we do:

```
Text("Charmander").font(.system(size:50)).fontWeight(.bold)
```

180

To have our pokemon image be in a circle shape with a white background, we do:

...

```
Image("charmander").resizable().padding().frame(height: 350.0)
  .background(Color.white).clipShape(Circle())
```

...

clipShape allows us to pass in a shape to mask our image.

Next, we add a shadow with a 10 point radius:

```
Image("charmander").resizable().padding().frame(height: 350.0)
  .background(Color.white).clipShape(Circle())
  .shadow(radius: 10 )
```

Figure 21

Now, we want our Charmander image to be shifted up with the fire type image to be in the background (fig. 22). To do so, we specify *offset* to be:

```
Image("charmander").resizable().padding().frame(height: 350.0)
  .background(Color.white).clipShape(Circle())
  .shadow(radius: 10 )
  .offset(x:0,y:-160)
```

Fig. 22

To move the image down, we specify a positive value for *y*. To move it up, we specify a negative value, i.e. -160.

Now we currently have quite a bit of space in between the pokemon image and the name text. To bring the name text up (fig. 23), we reduce the bottom padding of the image with:

```
Image("charmander").resizable().padding().frame(height: 350.0)
            .background(Color.white).clipShape(Circle())
            .shadow(radius: 10 )
            .offset(x:0,y:-160)
            .padding(.bottom,-150)
```

Figure 23

Reusable Views

For the next portion of our app, we are going to add some stats for our pokemon. We want to list each pokemon's hp, attack and defense stats in the same pattern (fig. 24).

Figure 24

And we want to make use of a reusable view that we could repeat three times. In that way, if we ever decided to change the font or size for the text, we could make that change once and it will reflect in each of the three stat layouts.

So go ahead and create our first custom view in SwiftUI. In Xcode, go to *File*, *New*, *File*, under 'iOS', select 'SwiftUI View' (fig. 25).

Figure 25

Name it *PokeStat* to describe its purpose. Click 'Next' and you will have a new custom view with a text view as a start.

We're going to need a horizontal stack to place an image view and two text views side by side for each stat row (fig. 26).

Figure 26

A quick way to embed the existing *Text* control in a *HStack* is to Command-click on *Text*, then select 'Embed in HStack' (fig. 27).

Figure 27

The code to envelop the *Text* control in a *HStack* will then be generated for you:

```
var body: some View {
    HStack {
        Text("Hello World!")
    }
}
```

You could of course type *HStack* with the curly brackets manually but it's convenient to use this especially when you do not know the specific code well enough.

Next, we add in our two text views:

```
struct PokeStat: View {
    var body: some View {
        VStack{
            HStack {
                Text("HP:")
                Text("39")
            }
        }
    }
}
```

To push it to the left, we add a *spacer* and also padding before the text:

```
HStack {

    Text("HP:").font(.system(size:40)).fontWeight(.bold).padding(.leading,30)
    Text("39").font(.system(size:40))
    Spacer()
}
```

Figure 28

Now that we've made this new stat text (fig. 28), how do we make use of it in our pokemon detail page?

Back in *ContentView* at the bottom, create a new stat text:

```
var body: some View {
    VStack{
        Image("firetype").resizable().padding().frame(height: 400.0)
        Image("charmander").resizable().padding().frame(height:
350.0)
            .background(Color.white)
            .clipShape(Circle())
            .shadow(radius: 10 )
            .offset(x:0,y:-160)
            .padding(.bottom,-150)

        Text("Charmander").font(.system(size:50)).fontWeight(.bold)
```

```
        PokeStat()
        Spacer()
    }.edgesIgnoringSafeArea(.top)
}
```

In our preview, our stat should show up like in figure 29.

Figure 29

Now, copy and paste to have three lines of *PokeStat()* (fig. 30).

Charmander
HP: 39
HP: 39
HP: 39

Figure 30

For now, our *PokeStat* contains hard coded data of "HP:39". We want our PokeStat to present dynamic data of course. To do so, in our struct *PokeStat*, add the following properties:

```
struct PokeStat: View {
```

```
    var statName: String
    var statValue: String
    ...
```

Now when you specify *PokeStat* in *ContentView.swift*, it will prompt you to enter values for *statName* and *statValue* i.e.:

```
PokeStat(statName:String, statValue: String)
```

Because we have added *statName* and *statValue* to *PokeStat*, the Preview function might be throwing some errors. To fix it, add the following:

```
struct PokeStat_Previews: PreviewProvider {
    static var previews: some View {
        PokeStat(statName: "", statValue: "")
    }
}
```

And in *ContentView.swift*, change the code to:

```
...
        VStack{
            ...
            Text("Charmander").font(.system(size:50)).fontWeight(.bold)

            PokeStat(statName: "HP", statValue: "39")
            PokeStat(statName: "Attack", statValue: "52")
            PokeStat(statName: "Defense", statValue: "43")

            Spacer()
        }.edgesIgnoringSafeArea(.top)
```

And in *PokeStat.swift*, to get those pieces of information using variables, make the change below:

```
struct PokeStat: View {

    var statName: String
    var statValue: String

    var body: some View {
        VStack{
            HStack {
                Text(statName)
                    .font(.system(size:40))
                    .fontWeight(.bold)
                    .padding(.leading,30)
```

```
            Text(statValue).font(.system(size:40))
            Spacer()
        }
      }
    }
}
```

You should finally get this screen like in figure 31.

Figure 31

Hopefully, you get the idea behind using reusable views in this section. We've created *PokeStat* to have its own properties that we can pass values into. This saves us from having to rewrite the same code three times.

SF Symbols

We are going to add symbols to our stats (fig. 32).

Figure 32

For that, we are going to use SF Symbols. SF symbols are included Xcode 11 and only applicable to iOS 13 devices and onwards. As stated in the Apple website (https://developer.apple.com/design/human-interface-guidelines/sf-symbols/):

"SF Symbols provides a set of over 2,400 consistent, highly configurable symbols you can use in your app....SF Symbols are available in a wide range of weights and scales to help you create adaptable designs."

SF Symbols

SF Symbols provides a set of over 2,400 consistent, highly configurable symbols you can use in your app. Apple designed SF Symbols to integrate seamlessly with the San Francisco system font, so the symbols automatically ensure optical vertical alignment with text in all weights and sizes.

Figure 33

To browse the full set of symbols, download the SF Symbols app which contains over 2,400 different symbols (fig. 34). (https://developer.apple.com/design/downloads/SF-Symbols.dmg).

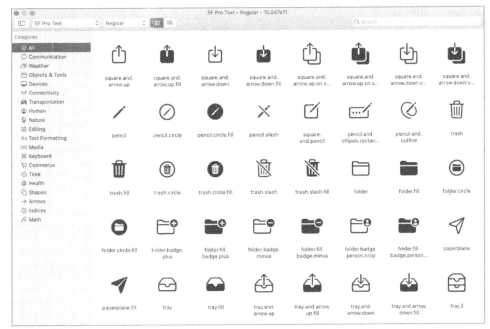

Figure 34

You can even create your own custom symbols as stated in Apple's website (fig. 35) But that is out of the scope of this book.

Creating Custom Symbols

If you need a symbol that isn't provided by SF Symbols, you can create your own. The SF Symbols app lets you export a symbol as a template in a reusable, vector-based file format. To create a custom symbol, export an SF symbol that's similar to the design you want and modify the template using a vector-editing tool like Sketch or Illustrator. Use the result in your app as you would use the original template file. (Custom symbols don't support adaptive color.) For developer guidance, see Creating Custom Symbol Images for Your App. See Symbols for Use As-Is for a list of symbols that can't be customized.

Be guided by the template. Create a custom symbol that's consistent with the system-provided ones in terms of level of detail, optical weight, alignment, position, and perspective. Strive to design a symbol that is:

- Simple
- Recognizable
- Not offensive
- Directly related to the action or content it represents

Figure 35

We will now show how to use SF symbols in our Pokemon app. To display an image containing a SF symbol, simply specify the *systemName* attribute for e.g.:

```
Image(systemName:"heart")
```

To add it into our Pokemon app, we specify another variable to store the SF symbol we want to display. In *PokeStat*, add the following:

```
struct PokeStat: View {

    var statName: String
    var statValue: String
    var statSymbol: String

    var body: some View {
        VStack{
            HStack {
                Image(systemName:statSymbol)
                    .font(.system(size:40))
                    .padding(.leading,30)
                Text(statName).font(.system(size:40)).fontWeight(.bold)
                Text(statValue).font(.system(size:40))
                Spacer()
            }
        }
    }
}
```

In previews, you will have to add:

```
    static var previews: some View {
        PokeStat(statName: "", statValue: "", statSymbol: "")
    }
```

And in *ContentView*, add:

```
...
    PokeStat(statName: "HP", statValue: "39", statSymbol: "heart")
    PokeStat(statName: "Attack", statValue: "52", statSymbol: "star")
    PokeStat(statName: "Defense", statValue: "43", statSymbol: "shield")
...
```

We would now have the SF symbols displayed (fig. 36):

192

Figure 36

Go ahead and try out other SF symbols! For each symbol, there is generally a normal one and a filled one (fig. 37).

Figure 37

Other places where SF symbols would serve to be useful are for example tab bar icons. So go ahead and give them a try!

Rows

In this next portion of our app, we are going to list out all the different pokemons in what used to be a *table view*, but is now called a *list* (fig. 38).

Figure 38

For this list to happen, we have to make our own custom row which consists of our pokemon picture and their name listed out.

We will create our own custom row view just as we did previously when we created *PokeStat*. Only this time, we're going to make it for one of the rows inside our list.

So, go to *File*, *New*, *File*, 'SwiftUI View', and call it *PokemonRow*.

Now, instead of showing the preview for the entire phone screen, we are going to simulate the preview of a single row with the below code:

```
struct PokemonRow_Previews: PreviewProvider {
    static var previews: some View {
        PokemonRow().previewLayout(.fixed(width: 500, height: 70))
    }
}
```

We've changed the preview to be similar to a row (fig. 39) so that we can better visualize how a row is going to look like.

Hello, World!

Figure 39

For the row, we have our image, text and a spacer to push it to the left with the following code:

```
struct PokemonRow: View {
    var body: some View {
      HStack{
        Image("charmander").resizable().frame(width:70, height:70)
        Text("Charmander").font(.largeTitle)
        Spacer()
      }
    }
}
```

And this is how our row looks (fig. 40):

 Charmander

Figure 40

Showing Multiple Rows in Preview

Now, showing a single row in the preview doesn't give a very good idea of how a list of rows is going to look like. It would be ideal if we could see how multiple rows look like in our preview.

To show multiple rows in preview, in the *PreviewProvider*, duplicate *PokemonRow* and enclose them in a *VStack*:

```
struct PokemonRow_Previews: PreviewProvider {
    static var previews: some View {
        VStack{
            PokemonRow().previewLayout(.fixed(width: 500, height: 70))
            PokemonRow().previewLayout(.fixed(width: 500, height: 70))
            PokemonRow().previewLayout(.fixed(width: 500, height: 70))
        }
    }
}
```

Our preview now shows multiple rows (fig. 41). Which gives a better idea of how multiple rows are displayed.

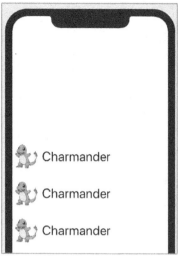

Figure 41

So you see how we can manipulate the preview to show us what suits our UI development best? This is a great feature of SwiftUI.

Currently, we have the same row duplicated. What if we want unique data for each row? To illustrate this, we will provide some dummy data.

Creating Dummy Data for Rows

We will store our dummy data in a separate file *Pokemon.swift*. To create the file, go to *File, New, File* and choose 'Swift File' (fig. 42).

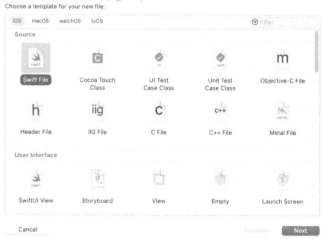

Figure 42

196

Click 'Next' and call it *Pokemon*.

In it, create a *struct Pokemon* with the following code:

```
import Foundation

struct Pokemon: Identifiable{
    var id: Int
    var name: String
    var imageName: String
    var type: String
    var hp: Int
    var attack: Int
    var defense: Int
}
```

Note that the struct adheres to the 'Identifiable' protocol, which means that instances of this struct can be uniquely identified via the *id* field.

Next, create an array *pokemons* that contains some dummy *Pokemon* instances:

```
let pokemons = [
        Pokemon(id: 0, name:"Charmander", imageName:"charmander",
            type:"firetype",hp:39,attack:52, defense:43),
        Pokemon(id: 1, name:"Bulbasaur",imageName:"bulbasaur",
            type:"grasstype",hp:45,attack:34, defense:43),
        Pokemon(id: 2, name:"Squirtle",imageName:"squirtle",
            type:"watertype",hp:45,attack:43, defense:45),
        Pokemon(id: 3, name:"Pikachu",imageName:"pikachu",
            type:"electrictype",hp:45,attack:46, defense:32),
        Pokemon(id: 4, name:"Machop",imageName:"machop",
        type:"fightingtype",hp:45,attack:43, defense:45)]
```

If you are interested, the pokemon stats are taken from https://pokemondb.net/pokedex/all.

* Do note that you will have to download the images of the above pokemon (and also the type images e.g. grass type, fighting type, water type), name them accordingly and place them in *Assets.cxassets* just as what we have done previously with the Charmander image (fig 43). Alternatively, you can use your own images and just change the image name in your code.

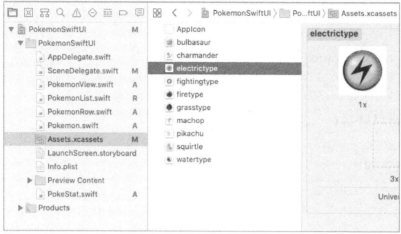

Figure 43

Having the list of dummy data, go back to *PokemonRow.swift* and make the following code changes:

```
struct PokemonRow: View {

    var pokemon: Pokemon

    var body: some View {
      HStack{
        Image(pokemon.imageName).resizable().frame(width:70, height:70)
        Text(pokemon.name).font(.largeTitle)
        Spacer()
      }
    }
}
```

Code Explanation

```
    var pokemon: Pokemon
```

First in *PokemonRow*, we create a variable *pokemon* to hold the *Pokemon* instance the row is representing.

```
    HStack{
      Image(pokemon.imageName).resizable().frame(width:70, height:70)
      Text(pokemon.name).font(.largeTitle)
      Spacer()
    }
```

Then in *HStack*, instead of hardcoding data, we replace it with *pokemon.imageName* and *pokemon.name*.

And in *PreviewProvider*, we show five rows of our data with:

```
struct PokemonRow_Previews: PreviewProvider {
    static var previews: some View {
        VStack{
            PokemonRow(pokemon:pokemons[0]).previewLayout(.fixed(width: 500,
                height: 70))
            PokemonRow(pokemon:pokemons[1]).previewLayout(.fixed(width: 500,
                height: 70))
            PokemonRow(pokemon:pokemons[2]).previewLayout(.fixed(width: 500,
                height: 70))
            PokemonRow(pokemon:pokemons[3]).previewLayout(.fixed(width: 500,
                height: 70))
            PokemonRow(pokemon:pokemons[4]).previewLayout(.fixed(width: 500,
                height: 70))
        }
    }
}
```

In the above code, we access the *pokemons* array in *Pokemon.swift*, and provide each element into *PokemonRow*.

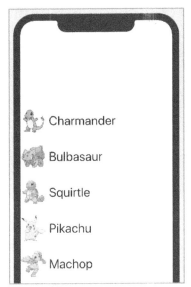

Figure 44

In the preview now, you should now get something like in figure 44, where you have data from *pokemons* array for each row. See how useful the preview allows us to view multiple rows of data at once without even running the simulator?

199

Now of course, we are now just using dummy data, but the *pokemons* array data could be populated via a JSON API call for example. Now that we have our row created, we're going to talk about how to create a list in the next section.

Lists

So far, we have a makeshift preview list in *PokemonRow*. To have an actual list view, create a new SwiftUI view file and call it *PokemonList*. We will show how simple it is to create a list in SwiftUI (fig. 45).

Figure 45

In *PokemonList.swift*, add in the following code:

```
struct PokemonList: View {
    var body: some View {
        List(pokemons){
            pokemon in PokemonRow(pokemon: pokemon)
        }
    }
}
```

List is a container that accepts an array and presents the array elements as rows of data arranged in a single column.

Next, we want to be able to tap on a row and navigate to the pokemon details page. To do so, we embed our *List* in a *NavigationView* and have each row embedded in a *NavigationLink*:

```
struct PokemonList: View {
    var body: some View {
        NavigationView{
```

```
List(pokemons){ pokemon in
    NavigationLink(destination: ContentView()){
        PokemonRow(pokemon: pokemon)
    }
}
}
}
```

In *NavigationLink*, we specify the destination it will go to when a user tap on it. In our case, it is *ContentView* which holds the pokemon's specific information. Note that for *NavigationLink* to work, it has to be in a *NavigationView*.

Previewing Our List

A powerful function of preview here is that when we hit the play button, we can get a live preview (fig. 46) and you could actually click on a row, navigate to the details page and also go back to the list page.

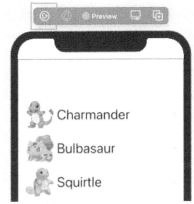

Figure 46

And we are doing all this without running the simulator which makes development and testing much more efficient.

Now let's add a 'Favorite Pokemons' title in our navigation bar (fig. 47).

Figure 47

Add to your *List* the code below:

```
...

        NavigationView{
            List(pokemons){ pokemon in
                NavigationLink(destination: ContentView()){
                    PokemonRow(pokemon: pokemon)
                }
            }
            .navigationBarTitle(Text("Favorite Pokemons"))
        }

...
```

Currently, each row navigates to the same details page. In the next section, we will work on passing the actual pokemon information for that specific row and get that to show in the details view.

Passing Data

We have been passing the pokemon information to each pokemon row. Now how do we pass that into the pokemon details view?

Firstly, our pokemon details view is named *ContentView* which is not very appropriate. Let's rename it from *ContentView* to *PokemonView*.

So in your project directory, rename *ContentView.swift* to *PokemonView.swift*.

In *PokemonView.swift*, rename *ContentView* to *PokemonView* as shown:

```
struct PokemonView: View {
    var body: some View {
        VStack{
        ...
        }.edgesIgnoringSafeArea(.top)
    }
}

struct PokemonView_Previews: PreviewProvider {
    static var previews: some View {
        PokemonView()
    }
}
```

Do the same in *PokemonList.swift* under *NavigationLink*:

...

```
                NavigationLink(destination: PokemonView()){
                    PokemonRow(pokemon: pokemon)
                }
```

...

When you do so, you might be prompted with an error in *SceneDelegate.swift* that you have to rename *ContentView* as shown below:

...
```
let contentView = ContentView()
```
...

However, instead of replacing it with *PokemonView()*, we replace it with *PokemonList()*. Because this view is the default root starting view.

Next in *PokemonView.swift*, add the following:

```
struct PokemonView: View {

    var pokemon: Pokemon

    var body: some View {
        VStack{
            Image(pokemon.type).resizable().frame(height: 400)
            Image(pokemon.imageName).resizable()
                .frame(height: 350.0).clipShape(Circle())
                .shadow(radius: 10)
                .offset(x:0,y:-160)
```

```
            .padding(.bottom,-150)

        Text(pokemon.name).font(.system(size:50)).fontWeight(.bold)

        PokeStat(statName: "HP", statValue: String(pokemon.hp),
            statSymbol: "heart")
        PokeStat(statName: "Attack", statValue:
            String(pokemon.attack), statSymbol: "star")
        PokeStat(statName: "Defense", statValue:
            String(pokemon.defense), statSymbol: "shield")

        Spacer()
    }.edgesIgnoringSafeArea(.top)
    }
}

struct PokemonView_Previews: PreviewProvider {
    static var previews: some View {
        PokemonView(pokemon:pokemons[0])
    }
}
```

Code Explanation

```
    var pokemon: Pokemon
```

We have a variable *pokemon* to hold the Pokemon object that will be passed in from the List.

```
        Image(pokemon.type).resizable()…
        Image(pokemon.imageName)….
            …
        Text(pokemon.name).font(…)
```

With the *pokemon* variable, we can than go ahead to populate the *type, imageName, name, hp, attack* and *defense* fields.

```
        PokeStat(statName: "HP", statValue: String(pokemon.hp),
            statSymbol: "heart")
        PokeStat(statName: "Attack", statValue:
            String(pokemon.attack), statSymbol: "star")
        PokeStat(statName: "Defense", statValue:
            String(pokemon.defense), statSymbol: "shield")
```

Take note that *hp, attack* and *defense* are integers. But because *PokeStat statValue* receives a *String*, we have to convert *hp, attack* and *defense* to *String* using *String(…)*.

```
struct PokemonView_Previews: PreviewProvider {
    static var previews: some View {
        PokemonView(pokemon:pokemons[0])
    }
}
```

And because *PokemonView* now takes in a *pokemon* variable, in our preview method, we simply provide the first pokemon element from our *pokemons* array.

Passing *pokemon* into *PokemonView*

Next, in *PokemonList.swift*, pass *pokemon* into *PokemonView*.

```
...
    var body: some View {
        NavigationView{
            List(pokemons){ pokemon in
                NavigationLink(destination: PokemonView(pokemon:
                    pokemon)){
                    PokemonRow(pokemon: pokemon)
                }
            }
            .navigationBarTitle(Text("Favorite Pokemons"))
        }
    }
    ...
```

Running your App

When you run your app and tap on a row, you should be able to see its Pokemon's detail page with its own statistics (fig. 48).

Figure 48

Congratulations! You've created your very first app with the SwiftUI framework.

A key functionality that aided our development process was the preview. This is especially so for detail views because in the past, we had to constantly run the simulator and do a series of taps from the main list view down to the detail view. But now, preview allows us to straight away see how our detail view looks like which makes development much faster.

There is of course a lot more to go through using SwiftUI for development. But as of now, this chapter serves as an introduction. In future, I might be releasing a separate book wholly dedicated to development using SwiftUI. If you would like to be notified when this book is released, just drop me a mail at support@i-ducate.com.

Chapter 12: Widgets

Widgets are one of the coolest things to come to iOS 14. People are using widgets to customize their home screens, to access essential and latest details from their apps at a glance (fig. 1).

Figure 1

So, if you can provide a nice widget to your app, users will love it. They have extra chances to see the latest information from your app in the home screen directly making it very convenient. If they tap on the widget, it takes them to the appropriate place in your app and thus use your app more.

We will walk you through creating a widget through one of your existing apps (PokemonUI chapter 11). We will have a widget that shows a new random pokemon's image and name hourly (fig. 2).

Figure 2

This will give you an intro into widgets that you can then use as a template for your apps.

Implementing our Widget

We first need to add a widget extension to the app. In the existing project, under Targets, click '+' (fig. 3):

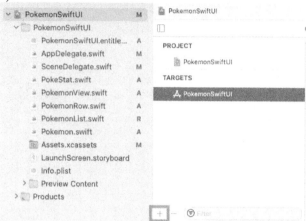

Figure 3

Under 'iOS', choose 'Widget Extension' and click 'Next' (fig. 4).

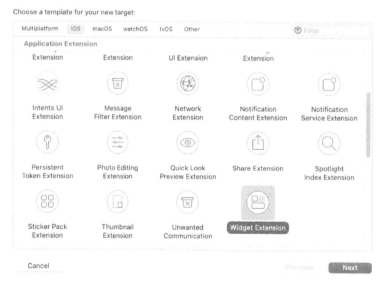

Figure 4

Name it 'PokemonWidget' and select 'Finish'.

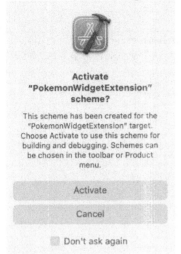

Figure 5

When you see the pop up (fig. 5), select 'Activate', and 'PokemonWidgetExtension' will then be added to your project. You will then be able to test run 'PokemonWidgetExtension':

Ⓔ PokemonWidgetExtension ⟩ 📱 iPhone 11

If we run it, it puts our widget on the home screen (fig. 6).

Figure 6

We have a widget prebuilt for us by Apple that simply shows the time (though we clearly need to improve on this).

Back in Xcode, you will notice that a new folder, 'PokemonWidget' has been created for you. In it, it has its own *Assets.xcassets* (where you can add images, colors, etc.), *Info.plist*, *PokemonWidget.intentdefinition* and *PokemonWidget.swift* (fig. 7).

Figure 7

If we look at *PokemonWidget.swift*, Xcode generates all these code for you. The code provides you with a template to control the widget much like the boilerplate code in a new table view controller.

SimpleEntry Struct

Let's first look at the *SimpleEntry* struct.

```
struct SimpleEntry: TimelineEntry {
    let date: Date
    let configuration: ConfigurationIntent
}
```

The *SimpleEntry* struct defines the data you provide for the widget. For example, in our current *SimpleEntry*, we define a *date* property for the widget.

212

You can see in *PokemonWidget.swift* that *SimpleEntry* is instantiated several times e.g.:

```
SimpleEntry(date: Date(), configuration: ConfigurationIntent())
```

Suppose we want the widget to also contain the data for a pokemon. We then add to *SimpleEntry*:

```
struct SimpleEntry: TimelineEntry {
    let date: Date
    let pokemon: Pokemon
    let configuration: ConfigurationIntent
}
```

But currently, Xcode will throw an error saying that it cannot find 'Pokemon' in scope. This is because *Pokemon.swift* is only in the *PokemonSwiftUI* target. To include *Pokemon.swift* in the 'PokemonWidgetExtension' target, select *Pokemon.swift*, under 'Target Membership', check on the 'PokemonWidgetExtension' checkbox (fig. 8).

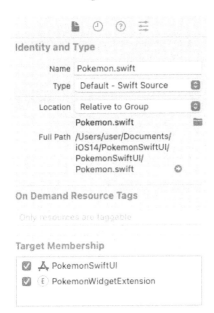

Figure 8

Struct SimpleEntry will then be able to reference 'Pokemon'. (you might need to 'Clean Build Folder' and 'Build' the project again)

And because we added a new property to *SimpleEntry*, we also have to provide a value for it when we instantiate *SimpleEntry*. To do so, add the below to all *SimpleEntry* instantiations:

```
SimpleEntry(date: Date(), pokemon: pokemons.randomElement()!,
configuration: ConfigurationIntent())
```

That is, we provide a random pokemon element to each *SimpleEntry* instantiation. You can see that *SimpleEntry* is instantiated in *placeholder*, *getSnapshot* and *getTimeline*:

```
    func placeholder(in context: Context) -> SimpleEntry {
        SimpleEntry(date: Date(), pokemon: pokemons.randomElement()!,
            configuration: ConfigurationIntent())
    }

    func getSnapshot(for configuration: ConfigurationIntent, in context:
Context, completion: @escaping (SimpleEntry) -> ()) {
        let entry = SimpleEntry(date: Date(),
            pokemon:pokemons.randomElement()!,
            configuration: configuration)
        completion(entry)
  }

    func getTimeline(for configuration: ConfigurationIntent, in context:
Context, completion: @escaping (Timeline<Entry>) -> ()) {
        var entries: [SimpleEntry] = []

        // Generate a timeline consisting of five entries an hour apart,
        // starting from the current date.
        let currentDate = Date()
        for hourOffset in 0 ..< 5 {
            let entryDate = Calendar.current.date(byAdding: .hour, value:
                hourOffset, to: currentDate)!
            let entry = SimpleEntry(date: entryDate,
                pokemon:pokemons.randomElement()!,
                configuration: configuration)
            entries.append(entry)
        }

        let timeline = Timeline(entries: entries, policy: .atEnd)
        completion(timeline)
   }
```

And also in the preview:

```
struct PokemonWidget_Previews: PreviewProvider {
    static var previews: some View {
        PokemonWidgetEntryView(entry: SimpleEntry(date: Date(),
            pokemon: pokemons.randomElement()!,
            configuration: ConfigurationIntent())))
```

```
        .previewContext(WidgetPreviewContext(family: .systemSmall))
    }
}
```

We will later explain the purposes of the above methods. For now, the widget preview is just showing (fig. 9):

Figure 9

Let's see how we can improve the widget's visual appearance.

Controlling the Visual Side of the Widget

We control the visual side of the widget in *PokemonWidgetEntryView*:

```
struct PokemonWidgetEntryView : View {
    var entry: Provider.Entry

    var body: some View {
        Text(entry.date, style: .time)
    }
}
```

As you can see, *PokemonWidgetEntryView* uses SwiftUI views to display the widget's content. We can't use storyboard for widgets. In *PokemonWidgetEntryView*, fill in the codes:

```
struct PokemonWidgetEntryView : View {
    var entry: Provider.Entry

    var body: some View {
        VStack{
            Image(entry.pokemon.imageName)
                .resizable()
                .scaledToFit()
                .frame(height: 80)
```

215

```
Text(entry.pokemon.name)
        .font(.system(size:25))
        .fontWeight(.bold)
Text(entry.date, style: .time)
    }
  }
}
```

PokemonWidgetEntryView populates the widget's view with the data set in the *Provider.Entry*. *Provider.Entry* is the data source for our widget's view.

We are trying to display a pokemon image with its name below (fig. 10).

Figure 10

Note that for the pictures to appear, we have to go to *Assets.xcassets* (under *PokemonSwiftUI*), select the pictures, and make sure under 'Target Membership' we check 'PokemonWidgetExtension' so that we can reference them from *PokemonWidgetEntryView*. We can now then display a random pokemon in our nice cute widget.

When you run your app, you will see your widget display a pokemon's image and name (fig. 11).

Figure 11

216

And when you click on the widget, it goes into your app!

Now let's revisit the rest of the code in *PokemonWidget.swift*.

getTimeline()

getTimeline() generates an array of *Timeline* entry objects. Each timeline entry specifies the date and time to update the widget's content. That is, you can specify when you want the widget to be updated. This is useful for example, a widget that displays weather information might update the temperature hourly throughout the day.

```swift
    func getTimeline(for configuration: ConfigurationIntent, in context: Context,
completion: @escaping (Timeline<Entry>) -> ()) {
        var entries: [SimpleEntry] = []

        // Generate a timeline consisting of five entries an hour apart, starting
        // from the current date.
        let currentDate = Date()
        for hourOffset in 0 ..< 5 {
            let entryDate = Calendar.current.date(byAdding: .hour,
                    value: hourOffset, to: currentDate)!
            let entry = SimpleEntry(date: entryDate,
                    pokemon: pokemons.randomElement()!,
                    configuration: configuration)
            entries.append(entry)
        }

        let timeline = Timeline(entries: entries, policy: .atEnd)
        completion(timeline)
    }
```

For example, the above default *getTimeline* code updates the widget 5 times an hour apart, since in the *for hourOffset in 0 ..<5*, we will in effect have:

```swift
// timelineentry now
entryDate = Calendar.current.date(byAdding: .hour, value: 0, to: currentDate)!

// timelineentry one hour in the future
entryDate = Calendar.current.date(byAdding: .hour, value: 1, to: currentDate)!

// timelineentry two hour in the future
entryDate = Calendar.current.date(byAdding: .hour, value: 2, to: currentDate)!

// timelineentry three hour in the future
entryDate = Calendar.current.date(byAdding: .hour, value: 3, to: currentDate)!

// timelineentry four hour in the future
entryDate = Calendar.current.date(byAdding: .hour, value: 4, to: currentDate)!
```

Each time shows a random pokemon. You can change the number and frequency of timeline entries. For e.g. instead of showing a different pokemon every hour, you can make it every 15 minutes by changing:

```
for hourOffset in 0 ..< 5 {
    let entryDate = Calendar.current.date(byAdding: .minute,
        value: hourOffset * 15, to: currentDate)!
    let entry = SimpleEntry(date: entryDate,
        pokemon: pokemons.randomElement()!,
        configuration: configuration)
    entries.append(entry)
}
```

Timeline Reload Policy

The timeline instance also contains a *TimelineReloadPolicy*.

```
let timeline = Timeline(entries: entries, policy: .atEnd)
```

The system would use this policy to determine when to invoke *getTimeline()* again to load the next set of timeline entries. By default, the policy is set as *atEnd*. This means after the fifth *SimpleEntry* is displayed on the widget, the system would trigger *getTimeline()* for the next batch.

Alternatively, you can specify *after(Date:)* to set a specific date at which the next timeline is fetched.

getSnapshot

getSnapshot is used to immediately present a widget view while the rest of the data is being loaded into the widget.

```
    func getSnapshot(for configuration: ConfigurationIntent, in context:
Context, completion: @escaping (SimpleEntry) -> ()) {
        let entry = SimpleEntry(date: Date(),
            pokemon:pokemons.randomElement()!,
            configuration: configuration)
        completion(entry)
    }
```

Widget Display Name and Description

You can also change the display name and description of the widget in the *PokemonWidget* struct:

```
@main
struct PokemonWidget: Widget {
```

218

```
let kind: String = "PokemonWidget"

var body: some WidgetConfiguration {
    IntentConfiguration(kind: kind, intent: ConfigurationIntent.self,
        provider: Provider()) { entry in
        PokemonWidgetEntryView(entry: entry)
    }
    .configurationDisplayName("My Widget")
    .description("This is an example widget.")
    }
}
```

kind is an identifier used ot distinguish the widget from others in the WidgetCenter.

The widget configuration display name and description is shown in the 'edit' widget page (Widget Gallery) where users personalize/arrange widgets to fit their specific needs. To do so, in home screen, scroll all the way left to see the widgets page and then scroll down, select 'Edit', '+' (fig. 12).

Figure 12

Summary

Widgets provide users a handy way for quick access to information without requiring them to open the full app. With iOS 14, many applications will be looking to leverage the WidgetKit framework and come up with exciting widgets in the coming future.

219

Chapter 13: App Clips

App clips are a way to take a lightweight version of your app and share it with users (who don't have the full app installed) to have some of its functionality when and where they need it. For example, the App Store has a coffee shop app that allows users to order a drink, save favorite drinks, collect rewards and so on. But say a walk-in customer walks into the coffee shop but doesn't have the full app downloaded. At the coffee stand is an app clip code or QR code tag to scan that downloads a small app clip that only offers functionality to order a drink.

An app clip is a small version of your app. It has to be small (less than 10MB) so that it can be downloaded quickly. The app clip launches instantly and helps the user perform a task as quickly as possible. So, at the coffee shop, they use it to quickly pick what they want to buy.

App clips are not a full download of your app, but just a clip that lives temporarily on your phone. If you don't interact with it for a while, it will eventually be deleted. If users like the clip, there are little prompts to download the full app. App clips don't appear on the Home screen and users don't manage them the way they manage full apps.

App clips are good particularly when it is used in particular locations or situations. For e.g., an app clip for ordering food in a restaurant, or to book rides in a theme park. They provide a polished user experience that helps users solve a real-world task as quickly as possible.

Creating an App Clip

In the *PokemonSwiftUI* project from either the last chapter or from chapter 11, add a new extension for our app clip by adding a new target (fig. 1).

Figure 1

Under 'Application, select 'App Clip', click 'Next' and name the clip 'PokemonClip' (fig. 2).

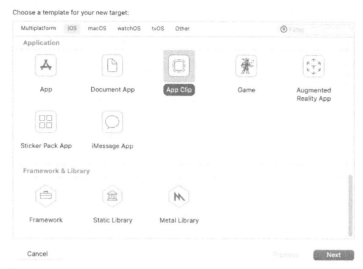

Figure 2

Under 'Interface', choose 'SwiftUI' and uncheck 'Core Data' (fig. 3).

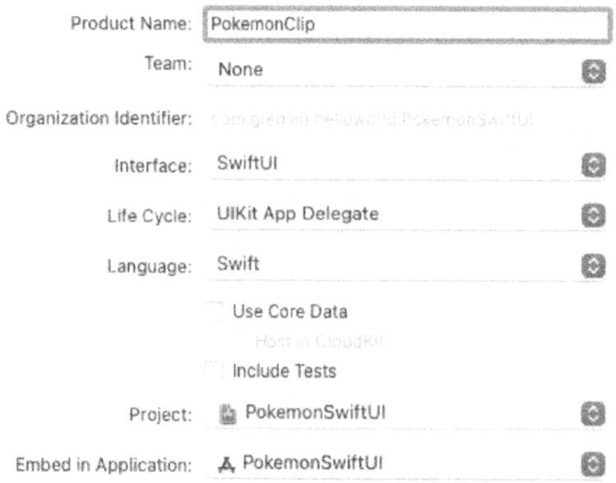

Figure 3

Select 'Finish' and activate 'PokemonClip' scheme.

There will a new folder called *PokemonClip* (fig. 4).

Figure 4

Notice that *PokemonClip* looks like a typical app having its own AppDelegate, ContentView etc. In fact, if your app is under 10 MB, your entire app can be an app clip!

Let's try to run PokemonClip on the simulator.

You will see the default 'Hello, world!' as we have not added anything to the clip's storyboard. But notice that the app clip's icon has dots around it. The dots signify that it's an app clip,

Remember that an app clip is just a test experience, and not the full app. The idea is we take only the essential parts of the full app and put it into the clip.
In our case, we will just show the list view listing the various pokemons. We will not be offering the function to navigate to each pokemon's details.

So, we can just copy the code from *PokemonList.swift* (under *PokemonSwiftUI*) and paste it into *ContentView.swift* (under *PokemonClip*). But we will remove the *NavigationView* and *NavigationLink* since we just want a list of the pokemons and remove the navigating to pokemon details functionality.

```
...
struct ContentView: View {
    var body: some View {
        List(pokemons){
            pokemon in PokemonRow(pokemon: pokemon)
        }
    }
}
...
```

There will be a few errors because we have not included *PokemonClip* under the target membership of some of the files.

So, select *Pokemon.swift* and under 'Target Membership', check 'PokemonClip' (fig. 5).

Figure 5

Do the same for *PokemonRow.swift* and also for the pokemon images, in *Assets.xcassets* (*PokemonSwiftUI*). We don't need to do for *PokemonView.swift* or *PokeStat.swift* since we are not using them.

When we run the app clip, it will have the reduced functionality of our existing app (just the list view). Other use case of clips is you could just include the beginning or some fun parts of your app.

As you can see, developing an app clip is pretty much the same as a full app. Clips use the same frameworks as full apps. Adding code or assets to an app clip works just like it does for any target. We can create new files and assets or use existing files or assets and add them as members to the app clip's target. To ensure the project's maintainability, the full app and the app clip should share as much code as possible.

There are many other stuff in iOS 14, but the two that make the difference to developers are widgets and app clips.

Chapter 14: Dark Mode

Because looking at screens with bright white lights can be a bit too harsh on the eyes (especially at night), Apple introduced dark mode in iOS 13.

Being integrated throughout iOS, dark mode re-skins the entire system, from Settings, Messages, Notes, to the on-screen keyboard and more (fig. 1).

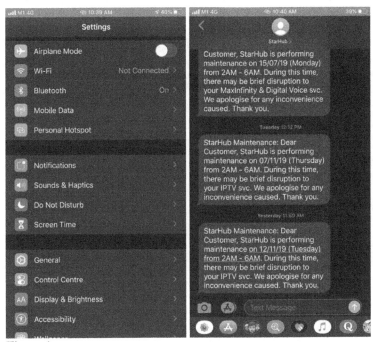

Figure 1

Now, if we have not set up our apps for dark mode, a user using dark mode might open our app and yet sees a view with bright lights because it has not been customized for dark mode. This potentially leads to bad user experience.

Another potential bad user experience if your app is not setup for dark mode is that Apple applies a default dark mode for you. For example, if you create a label and use the default color, it'll automatically work in dark mode. In dark mode, your text will be white and in light mode that text will be dark (fig. 2).

Label

Figure 2

But let's say I explicitly state that the label's text is black. When it transits to dark mode, because both background and text are black, the user can't see anything.

So, customizing your app for dark mode in iOS is an important thing you should have inside of your app. The good news is that if you use SwiftUI, a lot of it already caters to dark mode. For example, below are the views in dark mode for our Pokemon app developed in the previous chapter.

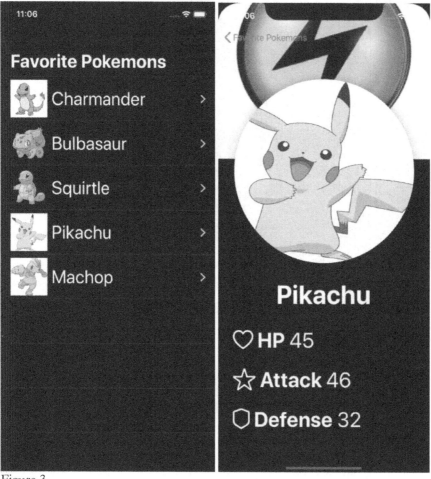

Figure 3

There is room for improvement though. For example, the background of the pokemon picture is still white. I should make it such that in dark mode, change to a darker background. We will show customizing of images between modes in a later section.

Dark Mode in Storyboard

Now, because most of our apps are still developed with storyboard, we will illustrate how to integrate dark mode into an app built using storyboard. In Xcode, create a new App project, call it 'DarkModeHelloWorld' and in 'User Interface', select 'Storyboard'.

In the storyboard, bring out like a label into the View Controller.

To see what your app looks like in either light or dark mode, under 'View as' in the bottom, select the 'Appearance' (fig. 4).

Figure 4

In dark mode, the background will go from white to black and the text does the opposite to white (fig. 5).

Figure 5

Now why is this so? If you select the top-level view and look at it's *Background* color, you can see its called 'System Background Color' (fig. 6)

Figure 6

227

'System Background Color' is white if someone's in light mode and black in dark mode. This is the standard default background color.

There are other variants of this. For example, if we change 'Background' to 'Secondary System Background Color', we get a slightly gray background in light mode.
In dark mode, it's a different shade of black. The secondary is supposed to create some distinction from the primary.

Go and explore on your own the different 'System' options that you can work with to give different shades of colors in between light and dark mode e.g. 'Tertiary System Background Color', 'Quaternary…'

We recommend using Apple's provided system colors as much as you can because they provide colors with optimal contrast in light and dark mode.

Custom Color

Other than using Apple's system colors, we can also create our own custom color for light and dark mode. To do so, go to 'Assets.xcassets' and click on the plus button at the bottom, select 'New Color Set' (fig. 7).

Figure 7

You can then set the color for 'Any Appearance' and 'Dark Apperance'. (fig. 8).

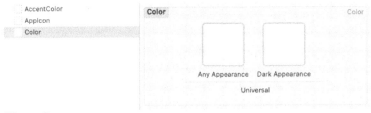

Figure 8

The 'Any Appearance' color on the left is for light mode and the 'Dark Appearance' on the right is for dark mode.

You can then specify different colors for the two modes using the sliding bars under 'Color' (fig. 9).

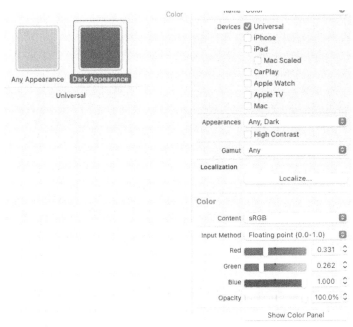

Figure 9

When you are done specifying the color, give the color set a custom name (fig. 10).

Color Set

Name CustomColor

Figure 10

Back in our storyboard, select the custom color that we have just created for any UI control (fig. 11):

Figure 11

The custom color will now be reflected when you toggle between light and dark mode (fig. 12).

Figure 12

Custom Images

We can also show one image in light mode and a different image in dark mode. To illustrate this, add an image view to our storyboard to display an image.

We create a custom image by selecting 'Assets.xcassets', click on the '+' button at the bottom and select 'Image Set'. Select the 'image', and in 'Appearances', specify 'Any, Dark'.

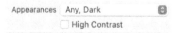

Next, drag the light mode image into the 'Any Appearance' dotted box and the dark mode image into the 'Dark Appearance' dotted box (fig. 13). I have chosen the image of a sun for light mode and a moon for dark mode.

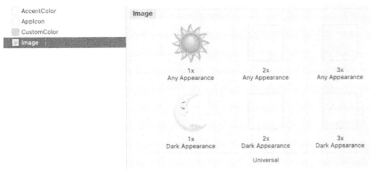

Figure 13

Back in 'Main.storyboard' under 'Image', select the image we have just created and it will show up (fig. 14).

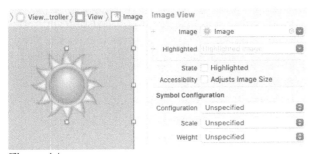

Figure 14

If you run the app in the simulator and toggle between light and dark mode now, it will show different images.

Figure 15

Summary

In this chapter, we learned how to integrate dark mode colors and images into our existing apps, thus providing users with a better experience. We explored using Apple's provided system colors to provide optimal contrast in light and dark mode, use custom colors and images and test how our interface looks in both appearances.

Chapter 15: Porting your iOS App to the Mac with Mac Catalyst

Mac Catalyst enables iOS developers to create MacOS apps from their iOS apps that they have already developed by just checking a tick box in Xcode. In doing so, we merge development for iOS and Mac apps. This makes it far simpler for developers to translate their apps across iPhone, iPad, or Mac. In this chapter, we will illustrate this by porting our Pokemon iOS app (from chapter 11) into a MacOS app.

First, open the project from chapter nine in Xcode. Next, go to the settings of the app, by clicking on project, and under *targets*, click on the 'Mac' checkbox (fig. 1).

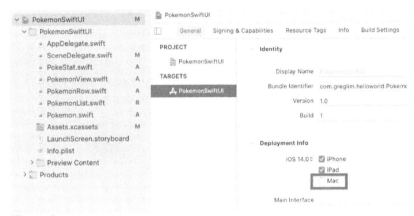

Figure 1

So click on the checkbox, and in the 'Enable Mac support' pop up box that appears next, click 'Enable' (fig. 2).

Figure 2

Next, we have to go to *signing and capabilities* because whenever we want to deploy an app to a physical device such as an iPhone, iPad or Mac, we have to provide a signing certificate.

So go ahead and sign in with your Apple I.D. You should be able to select your team in the team dropdown list (fig. 3).

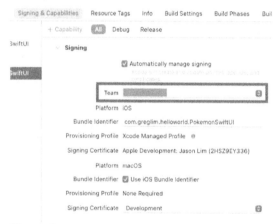

Figure 3

Now that we've signed our app and checked Mac capabilities, ensure that the Mac has been selected as the run target (fig. 4) and click *Run*.

PokemonSwiftUI ⟩ My Mac

Figure 4

*At time of writing, I encountered this error in *AppDelegate.swift* when I ran my app:
'*AppDelegate*' *is annotated with @main and must provide a main static function of type () -> Void or () throws -> Void.*
To rectify this, change @*main* to @*UIApplicationMain*:

```
import UIKit

//@main
@UIApplicationMain
class AppDelegate: UIResponder, UIApplicationDelegate {
...
```

When your app runs successfully, you should see the Mac version of your iOS app run (fig. 5).

Figure 5

Upon seeing this for the first time, I was totally amazed! In my past experience of developing Mac apps, it was such a pain. Now, it is just a matter of checking a box!

Now porting your iOS apps to a Mac app is also available for apps built using the storyboard. Project Catalyst actually ports UIKit over to the Mac so that iOS apps can be built and run on the Mac.

So in your storyboard projects, simply do the same steps as we have covered. That is to say, check the Mac check box, click 'Enable', and under 'Signing and Capabilities' select your development team. And click on 'Run'.

And once it runs, as long as you've got all of your constraints set up properly, it should try and fill the screen even though it is a different size and aspect ratio.

In summary, Mac Catalyst is a really cool piece of technology making it easy for us to port apps from iPhone, iPad to the Mac by just using a couple of checkboxes.

Chapter 16: In-App Purchases

In this chapter, we are going to learn about In-App Purchases and how to implement them for our apps. If you ever played games on the iPhone, you should be familiar with in-app purchases. For example, in Pokemon Go, you can buy pokecoins with in-app purchases. Another popular model is the freemium model, where an app gives a user a portion of the functionality for free, and then put up an in-app purchase to upgrade to the premium version for full functionality (or to remove ads).

In-app purchases work well. The top-grossing iPhone apps in the App Store are usually free to download but then include in-app purchases typically. Being free make it easier for users to download your app, giving it a go, and seeing what the features are before you start charging them for premium features or to remove ads. And that is how these apps make huge profits. Note that Apple charges you 30% of everything you make through in-app purchases.

Before we get started adding in-app purchases to our app, there are two things that you need before you can run or test the code. The first thing you need is a full Apple developer program. You have to enroll in the program before you can start adding in-app purchases to your apps in Xcode and start testing them. So, this means that you have to pay $99 per year.

The other requirement is that you need a physical iPhone device to test in-app purchases. The code for in-app purchases won't work on the simulator, and it doesn't tell you explicitly that it's not working because you're running on a simulator.

Let's now see what we'll build by the end of this chapter. Remember the Quick Quotes app in chapter two? We had a simple app that displays a table view of quotes (fig 1).

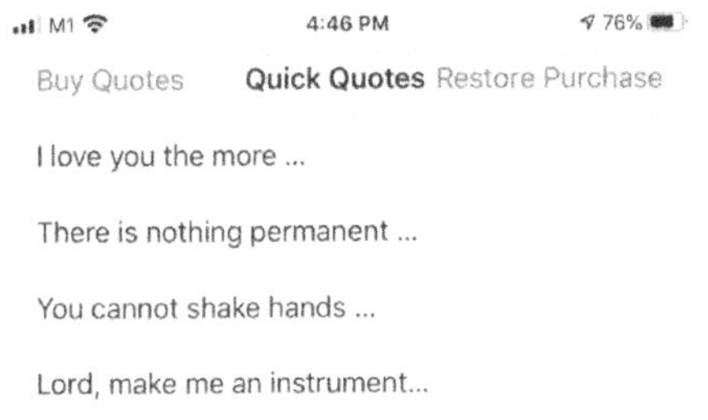

ᵃᵗᵗᵖ M1 📶	4:46 PM	⬥ 76% 🔋
Buy Quotes	**Quick Quotes** Restore Purchase	

I love you the more …

There is nothing permanent …

You cannot shake hands …

Lord, make me an instrument…

Figure 1

We will give the user access to the quotes in chapter three for free. But we will also have a group of paid quotes that will be displayed only when the user purchases an in-app purchase by tapping on the top left 'Buy Quotes' bar button item.

Although this is a simple app, in the process, we're going to look at how to make purchases, how to restore in-app purchases and how to detect if the purchase is successful or if there are errors. By the end of the chapter, you should understand how to implement in-app purchases for your own app.

Setting Up In-App Purchases on App Store Connect

First, go to https://developer.apple.com/account, 'Certificates, IDs & Profiles' (fig 2).

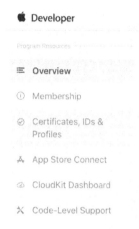

Figure 2

Under 'Identifiers', you will see a list of App IDs (fig. 3).

Figure 3

Click the '+' button, select 'App IDs' and 'App' to add a new one (fig. 4).

Register a new identifier

Select a type

Figure 4

It will take you through the process of registering a new app ID and register your app on Apple's central database. The name of our app is going to be 'QuickQuotes'.

Under 'Bundle ID', select the 'Explicit' checkbox. Then set the bundle ID. Remember that bundle ID has to be unique in Apple's database. If you have a business domain name, e.g. www.i-ducate.com, your bundle ID will be *com.i-ducate.quickquotes* (fig. 5).

Register an App ID

Platform	App ID Prefix
iOS, macOS, tvOS, watchOS	8Y8H59JVBJ (Team ID)
Description	Bundle ID ● Explicit ○ Wildcard
Show Quick Quotes	com.greglim.quickquotes
You cannot use special characters such as @, &, *, ', "	We recommend using a reverse-domain nar com.domainname.appname). It cannot cont

Figure 5

If you don't have a company website, just do *com.<firstnamelastname>* i.e. *com.greglim* and append your app name at the end (*com.greglim.quickquotes*). Note that the capitalizations matter!

An essential thing is that the bundle ID should be the **same** as our bundle identifier back in Xcode, under 'General' tab, 'Identity' (fig. 6).

General	Signing & Capabilities	Resource Tags	Info	Build

▼ Identity

ɹotes

Display Name QuickQuotes

ɹotes

Bundle Identifier com.greglim.quickquotes

Version 1.0

Build 1

Figure 6

Next, scroll down to the 'Capabilities' section and see that 'Game Center' and 'In-App Purchase' are ticked by default.

Figure 7

When you create a new app ID, they will auto-include these capabilities. Click on 'Continue', and the button will change to 'Register'. Select it and you are done. Your app info will begin to propagate through Apple's server and databases.

App Store Connect

Next, go to *appstoreconnect.connect.com* (fig. 8).

Figure 8

240

Agreements, Tax and Banking

In App Store Connect, you have to make sure that all your agreements, tax, and banking agreements are all ok. Check if you have any pending agreements that should be taken of, for example:

Review the updated Paid Applications Schedule.
In order to update your existing apps, create new in-app purchases, and submit new apps to the App Store, the user with the Legal role (Account Holder) must review and accept the Paid Applications Schedule (Schedule 2 to the Apple Developer Program License Agreement) in the Agreements, Tax, and Banking module.

To accept this agreement, they must have already accepted the latest version of the Apple Developer Program License Agreement in their account on the developer website.

This is a critical step that must not be overlooked. I spent quite a bit of time wondering why I couldn't get my in-app purchases to work in the Sandbox environment until I realized it is because I have not set up 'Tax, Banking and Contact Agreements' properly (fig. 9). So make sure you do so.

App Store Connect Agreements, Tax, and Banking ⌄

Jason Lim ⌄
Jason Lim (2)   (?)

Agreements   Tax   Banking   Contacts

Agreements

Type ⌄	Countries or Regions ⌄	Effective ⌄	Status ⌄	Action
Free Apps ?	All Countries or Regions View	Mar 29, 2020 - Mar 30, 2021	⦿ Active	
Paid Apps ?	All Countries or Regions View	Apr 6, 2020 - Mar 30, 2021	⦾ Processing	

Figure 9

New App

Back in the *App Store Connect* main menu, click on 'My Apps'. Click on the '+' button to create a new app. Fill in the form, along with your app name (fig. 10).

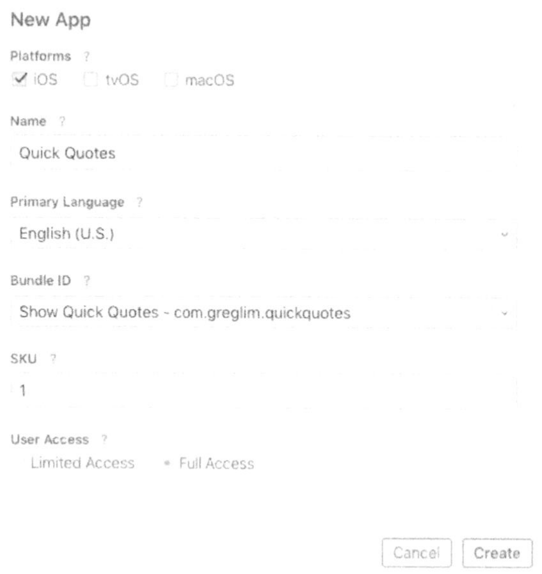

New App

Platforms ?

☑ iOS ☐ tvOS ☐ macOS

Name ?

Quick Quotes

Primary Language ?

English (U.S.) ⌄

Bundle ID ?

Show Quick Quotes - com.greglim.quickquotes ⌄

SKU ?

1

User Access ?

Limited Access   • Full Access

Cancel   Create

Figure 10

Note that your app name has to be unique in the App Store. If it is not, the form will prompt you to enter a unique name. For e.g. 'Quick Quotes' is already taken, so I use 'Quick Quotes SG' instead. ('SG' because I am a Singaporean)

In bundle ID, select the app's bundle ID that we have just added earlier. SKU (stock keeping unit) is a unique id you give to your apps to identify it among your own app catalog. The SKU is hidden from the App Store. When all is done, click 'Create'.

Creating an In-App Purchase

Next, go to the 'Features' tab. Under 'In-App Purchases', click on the '+' icon to create a new In-App purchase (fig. 11).

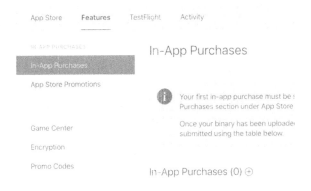

App Store   **Features**   TestFlight   Activity

IN-APP PURCHASES   In-App Purchases

In-App Purchases

App Store Promotions

Your first in-app purchase must be s
Purchases section under App Store

Once your binary has been uploader
submitted using the table below.

Game Center

Encryption

Promo Codes

In-App Purchases (0) ⊕

Figure 11

Several in-app purchase options are shown to you (fig 12):

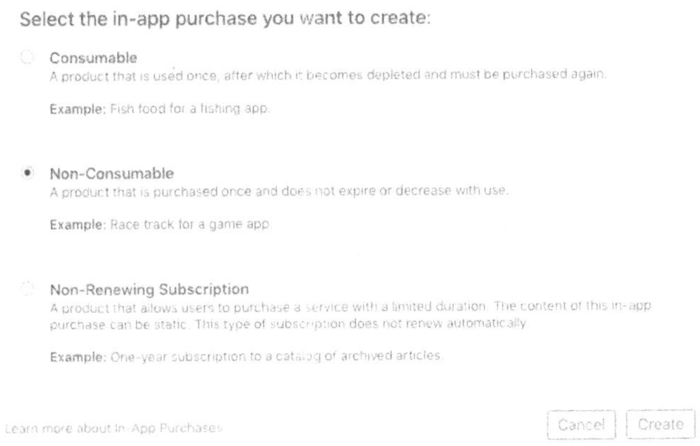

Select the in-app purchase you want to create:

○ **Consumable**
A product that is used once, after which it becomes depleted and must be purchased again.

Example: Fish food for a fishing app.

● **Non-Consumable**
A product that is purchased once and does not expire or decrease with use.

Example: Race track for a game app.

○ **Non-Renewing Subscription**
A product that allows users to purchase a service with a limited duration. The content of this in-app
purchase can be static. This type of subscription does not renew automatically.

Example: One-year subscription to a catalog of archived articles.

Learn more about In-App Purchases   Cancel   Create

Figure 12

As mentioned in the form, 'Consumable' is used commonly in games, e.g. poke coins, health.

'Non-Consumable' is purchased once e.g. getting rid of ads. The app will not ask the user for such a purchase again.

Lastly, we have 'Non-Renewable Subscription' e.g. a one year's subscription to a catalog of archived articles.

We will use the 'Non-Consumable' in-app purchase which is the most frequent in-app purchase type. So select that option and hit on 'Create'.

Fill in a reference name for your in-app purchase e.g. 'Paid Quotes'. It doesn't appear anywhere in your app but serves as just a reference for yourself. In 'Product ID' a good practice is to use your bundle ID + in-app purchase ID e.g. *com.greglim.quickquotes.paidquotes* (fig. 13).

Reference Name   ?

Paid Quotes

Product ID   ?

com.greglim.quickquotes.paidquotes

Availability   ?

☑ Cleared for Sale

Figure 13

Next, choose a price. E.g. USD 0.99 (Tier 1)

Under 'Localizations', fill in a display name which is seen by the user when they purchase the in-app purchase. E.g. 'Paid Quick Quotes'. You can choose to add other localizations e.g. Chinese.

Next, you have to add a screenshot of your in-app purchase in action. For now, just upload a random image for the screenshot, else there will be a 'Missing Metadata' warning and we can't proceed with developing our app. You can revisit this and upload a proper screenshot after you have created your app.

When you finish adding your in-app purchase, ensure that its 'Status' showing in App Store Connect is 'Ready to Submit' (fig. 14).

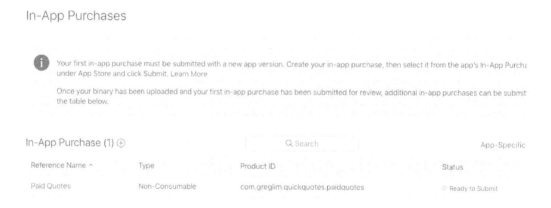

In-App Purchases

ⓘ Your first in-app purchase must be submitted with a new app version. Create your in-app purchase, then select it from the app's In-App Purcha under App Store and click Submit. Learn More

Once your binary has been uploaded and your first in-app purchase has been submitted for review, additional in-app purchases can be submit the table below.

In-App Purchase (1) ⊕   Q Search   App-Specific

Reference Name ^	Type	Product ID	Status
Paid Quotes	Non-Consumable	com.greglim.quickquotes.paidquotes	◌ Ready to Submit

Figure 14

Working on our Quick Quotes App

Back in Xcode, make sure that you have opened the QuickQuotes project from chapter two. If you do not have the project code, request the source code from me (support@i-ducate.com). Or if you have not gone through that chapter, please go back and review it first.

Now, ensure that in your Xcode QuickQuotes project, the bundle ID is the same one as you have created earlier in *developer.apple.com* under 'Identifier', 'App IDs'.

Adding In-App Purchase Capability

In Xcode, under 'Targets', under 'Signing & Capabilities', click on '+ Capability' and select 'In-App Purchase' (fig. 15, 16).

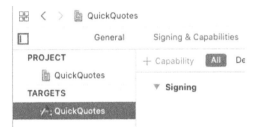

Figure 15

Figure 16

Making a Purchase Request

First, in our project from chapter two, on the top left of our Table View, add a bar button item that says, 'Buy Quotes' (fig. 17).

Figure 17

When the button is clicked, it will trigger the purchase.

Click drag on the 'Buy Quotes' bar button to *QuoteTableViewController.swift* and create an IBAction called *buyQuotes*:

```
@IBAction func buyQuotes(_ sender: Any) {
}
```

StoreKit is the framework which allows us to deal with in-app purchase related concepts. We have to import *StoreKit()* by adding:

```
import UIKit
import StoreKit
```

Then, implement *buyQuotes* as shown below:

```
@IBAction func buyQuotes(_ sender: Any) {
    if SKPaymentQueue.canMakePayments(){
        let paymentRequest = SKMutablePayment()
        paymentRequest.productIdentifier =
"com.greglim.quickquotes.paidquotes"
        SKPaymentQueue.default().add(paymentRequest)
    }
    else{
        print("user can't make payments")
    }
}
```

Code Explanation

```
    if SKPaymentQueue.canMakePayments(){
```

We first check if the user can purchase with *SKPaymentQueue.canMakePayments()*. For example, there might be parental controls to prevent kids from spending thousands of dollars in PokemonGo. *SKPaymentQueue* is the queue of payment transactions to be processed by the App Store.

```
if SKPaymentQueue.canMakePayments(){
    let paymentRequest = SKMutablePayment()
    paymentRequest.productIdentifier =
"com.greglim.quickquotes.paidquotes"
    SKPaymentQueue.default().add(paymentRequest)
}
```

If *canMakePayments* returns true, we go ahead to make a payment request by instantiating a *SKMutablePayment* object, *paymentRequest*. We set *paymentRequest*'s *productIdentifier* to the Product ID of the in-app purchase we created earlier on, i.e. "*com.greglim.quickquotes.paidquotes*" (fig. 18)

Reference Name ^	Type	Product ID
Paid Quotes	Non-Consumable	com.greglim.quickquotes.paidquotes

Figure 18

We then add *paymentRequest* to *SKPaymentQueue*.

Observing for a Successful Transaction

Now that we have added our payment request to the queue, we observe if the transaction succeeds or fails. We do this by implementing the *SKPaymentTransactionObserver* which notifies us when a payment transaction succeeds or fails:

```
import UIKit
import StoreKit

class QuoteTableViewController: UITableViewController,
SKPaymentTransactionObserver {
...
```

When you implement *SKPaymentTransactionObserver*, Xcode will prompt you to implement the *paymentQueue* delegate method to conform to its protocol.

```
class QuoteTableViewController: UITableViewController, SKPaymentTransactionObserver {

    var quotes = [
        "I love you the more
        "There is nothing pe
```

> ⊘ Type 'QuoteTableViewController' does not conform to protocol 'SKPaymentTransactionObserver'
>
> Do you want to add protocol stubs?   Fix

You can either click on 'Fix' to have Xcode automatically generate the method for you. Or you can type it out manually on your own as shown below:

```
func paymentQueue(_ queue: SKPaymentQueue, updatedTransactions
transactions: [SKPaymentTransaction]) {
      <#code#>
}
```

paymentQueue will be called when payment transactions are updated in the payment queue. In *paymentQueue* is the argument *transactions* which is an array of *SKPaymentTransactions* containing the updated transactions. The reason *transactions* is an array is because more than one payment request can be made at the same time.

To check if our transaction is successful, we loop through the *transactions* array and check the state of each transaction. We do by implementing *paymentQueue*:

```
func paymentQueue(_ queue: SKPaymentQueue, updatedTransactions
transactions: [SKPaymentTransaction]) {
      for transaction in transactions{
        if transaction.transactionState == .purchased{
          print ("transaction successful")
        } else if transaction.transactionState == .failed{
          print ("transaction failed")
        }
      }
}
```

Using a *for*-loop, we loop through *transactions* and check if the transaction state is purchased or failed. A transaction state can fail due to a user clicking on 'Cancel' and doesn't want to continue with the payment.

Next, we have to declare that our current class, *QuoteTableViewController* is the class receiving the notification messages from *SKPaymentTransactionObserver* when the transaction changes. To do so, add in *viewDidLoad()*:

```
override func viewDidLoad() {
    super.viewDidLoad()
    SKPaymentQueue.default().add(self)
}
```

248

Do note that we have previously also called *SKPaymentQueue.default().add*, when we added a payment request e.g. *SKPaymentQueue.default().add(paymentRequest)*.

```
Void add(observer: SKPaymentTransactionObserver)
Void add(payment: SKPayment)
```

This is an example of method *overloading*, where *add* has different implementations based on what argument it is receiving. *add* has a different implementation if it receives an argument of type *SKPaymentTransactionObserver* and another implementation when it receives an argument type of *SKPayment*. Now, we are using the *Void add(observer: SKPaymentTransactionObserver)* implementation.

Having implemented these, we are now ready to test our app on an actual device. Because we cannot be testing our app with actual money in the real App Store, we will be testing our app in a sandbox environment with sandbox users. Let's go ahead to create our sandbox users.

Creating Sandbox Users

Back in *appstoreconnect.apple.com*, go to 'Users and Access'. Under 'Sandbox', go to 'Testers' (fig. 19).

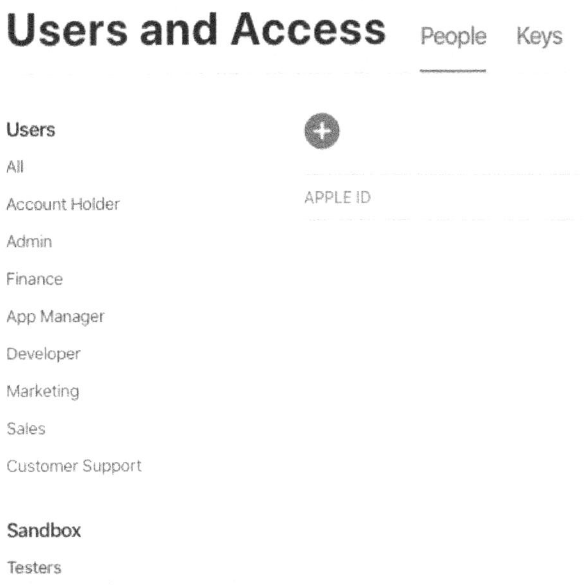

Users and Access People   Keys

Users

All

Account Holder   APPLE ID

Admin

Finance

App Manager

Developer

Marketing

Sales

Customer Support

Sandbox

Testers

Figure 19

This is where we add sandbox testers. Click on the '+' to add a new Tester. For each tester, you will have to use a unique email and this is quite a pain given that we don't usually have so many emails. To work around this, you can use temporary disposable emails for e.g. https://temp-mail.org/en/ to create the testers.

For the rest of the Tester Form fields, you can fill in fake information if you wish, except the 'App Store Country' field, where you have to choose the App Store your phone is currently signed in to.

Now, plug in your phone through USB, select the device as destination and run the app. Before we do anything, on the actual phone, go to 'Settings' and logout from your current iTunes account before trying to make any purchase.

When that is done, click on the 'Buy Quotes' bar button and when you are prompted to sign in, sign in with the test user you just created. Enter your password, click on 'buy', and you should have 'purchase successful' and printed in the console 'transaction successful'.

If your transaction fails

In the event that our transaction fails, we have to know the reason why. To do so, we add the following log statements in the transaction fail clause:

```
} else if transaction.transactionState == .failed{
   if let error = transaction.error{
       let errorDesc = error.localizedDescription
       print ("transaction failed due to error:\(errorDesc)")
   }
}
```

In the majority of cases, you will receive the message, "transaction failed due to error: **Cannot connect to iTunes Store**".

This indeed is a very difficult error to debug because Apple doesn't provide you the reason the transaction fails. But you can check for the following:
-   Have you logged out from your current App Store Connect account in your device before trying to make any purchase?
-   Have you enabled In-App Purchases for your App ID?
-   Is your in-app purchase's status in App Store Connect 'Ready to Submit'?
-   Does your project's .plist Bundle ID match your App ID?
-   Are you using the full in-app purchase product ID when making a *SKMutablePayment*?
-   Are your tax, bank and payment contracts approved in App Store Connect?
-   Is your device jailbroken? If so, you need to revert the jailbreak for IAP to work.

Giving Users Access to Purchased Content

Next, we will enable users to access whatever they have purchased once we get a successful transaction reported back to us from the payment queue. We first declare a new array called *paidQuotes* which contain the premium quotes that will be displayed when a user makes the purchase.

```swift
class QuoteTableViewController: UITableViewController,
SKPaymentTransactionObserver {

    var quotes = [
        "I love you the more …",
        "There is nothing permanent …",
        "You cannot shake hands …",
        "Lord, make me an instrument…"
    ]

    var paidQuotes = [
        "Love For All, Hatred For None.",
        "Every moment is a fresh beginning.",
        "Aspire to inspire before we expire",
        "Whatever you do, do it well."
    ]
    …
```

Next, in the 'purchased' *if*-clause, add the below codes:

```swift
    func paymentQueue(_ queue: SKPaymentQueue, updatedTransactions
transactions: [SKPaymentTransaction]) {
        for transaction in transactions{
           if transaction.transactionState == .purchased{
             quotes.append(contentsOf: paidQuotes)
             tableView.reloadData()

             navigationItem.setLeftBarButton(nil, animated: true)
             SKPaymentQueue.default().finishTransaction(transaction)
             print ("transaction successful")
           } else if transaction.transactionState == .failed{
                 if let error = transaction.error{
                let errorDesc = error.localizedDescription
                print ("transaction failed due to error:\(errorDesc)")
             }
          }
        }
    }
```

Code Explanation

```
quotes.append(contentsOf: paidQuotes)
tableView.reloadData()
```

We append the original quotes array with the contents of *paidQuotes*. We then reload the table view with *tableView.reloadData()*.

```
navigationItem.setLeftBarButton(nil, animated: true)
```

Because the purchase is made, we remove the 'Buy Quotes' button on the left of the bar.

```
SKPaymentQueue.default().finishTransaction(transaction)
```

Lastly, we finish the transaction.

Running your App

Run your app now and sign in with a new sandbox tester. Click the 'Buy Quotes' button and the paid quotes should appear rightly. The 'Buy Quotes' bar button item should also be removed.

For users who have already purchased the in-app purchase, but they for some reason, delete the app, get a new phone, or update to a new iOS version, their apps will be deleted. Will they have to re-purchase when they re-install the app? No, Apple will detect that they have purchased the content before and prompt them, "You've already purchased this. Would you like to get it again for free?" and when they click 'Ok', the in-app purchase will be restored.

But instead of asking such users them to click on the buy button, we should instead be asking them to click on a 'Restore Purchase' button. After a user has paid for a non-consumable product, she should never be asked to pay for it again. Apple is thus quite strict about providing users with the ability to restore in-app purchases they have bought before. We will implement this in the next section.

Restoring In-App Purchases

The 'Restore Purchase' button has to be prominently displayed and understandable by users. Apple wants users to get the best experience to restore IAP purchases conveniently. Without the option to restore, get ready to see your app being rejected from the App Store. Thus, we will add our 'Restore Purchase' bar button on the right of our Quick Quotes header (fig. 20).

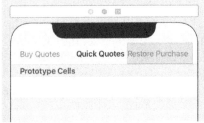

Figure 20

Create an *IBAction* for it and add in the below code to restore previous completed transactions:

```
@IBAction func restorePurchase(_ sender: Any) {
    SKPaymentQueue.default().restoreCompletedTransactions()
}
```

We then have to add to our payment queue:

```
func paymentQueue(_ queue: SKPaymentQueue, updatedTransactions
transactions: [SKPaymentTransaction]) {
    for transaction in transactions{
        if transaction.transactionState == .purchased{
            quotes.append(contentsOf: paidQuotes)
            tableView.reloadData()
            navigationItem.setLeftBarButton(nil, animated: true)
            navigationItem.setRightBarButton(nil, animated: true)

            SKPaymentQueue.default().finishTransaction(transaction)
            print ("transaction successful")
        } else if transaction.transactionState == .failed{
            if let error = transaction.error{
                let errorDesc = error.localizedDescription
                print ("transaction failed due to
error:\(errorDesc)")
            }
        }
        else if transaction.transactionState == .restored {
            quotes.append(contentsOf: paidQuotes)
            tableView.reloadData()
            navigationItem.setLeftBarButton(nil, animated: true)
            navigationItem.setRightBarButton(nil, animated: true)

            print("transaction restored")
            SKPaymentQueue.default().finishTransaction(transaction)
        }
    }
}
```

The main addition we did was to add the *if*-clause for the 'restored' transaction state. Just like for the 'purchased' state, we append *paidQuotes* to *quotes* and reload the table view. We then set the 'Buy Quotes' left bar button and also the 'Restore Purchase' right bar button to be disabled.

Additionally, back in 'purchased' state, we disable the 'Restore Purchase' bar button and thus add the line:

```
if transaction.transactionState == .purchased{
   quotes.append(contentsOf: paidQuotes)
   tableView.reloadData()
   navigationItem.setLeftBarButton(nil, animated: true)
   navigationItem.setRightBarButton(nil, animated: true)
   ...
```

Running your App

When you run your app now, you can tap on the restore button to restore your In-App Purchase.

Summary

We have gone through quite a lot of content to equip you with the skills to create an iOS app and submit it to the app store.

Hopefully, you have enjoyed this book and would like to learn more from me. I would love to get your feedback, learning what you liked and didn't for us to improve.

Please feel free to email me at support@i-ducate.com if you encounter any errors with your code or to get updated versions of this book.

If you didn't like the book, or if you feel that I should have covered certain additional topics, please email us to let us know. This book can only get better thanks to readers like you.

If you like the book, I would appreciate if you could leave us a review too. Thank you and all the best for your learning journey in iOS development!

About the Author

Greg Lim is a technologist and author of several programming books. Greg has many years in teaching programming in tertiary institutions and he places special emphasis on learning by doing.

Contact Greg at support@i-ducate.com or find out more at www.greglim.net